流域环境污染防控系列丛书

流域磷污染来源及源解析

张 涵 胡远思 著

西南交通大学出版社
·成 都·

图书在版编目（CIP）数据

流域磷污染来源及源解析 / 张涵，胡远思著.
成都：西南交通大学出版社，2024.8. -- ISBN 978-7
-5643-9945-0

Ⅰ．X522

中国国家版本馆 CIP 数据核字第 2024RT0738 号

Liuyu Lin Wuran Laiyuan ji Yuanjiexi
流域磷污染来源及源解析

张　涵　胡远思　著

策 划 编 辑	秦　薇
责 任 编 辑	赵永铭
封 面 设 计	GT 工作室
出 版 发 行	西南交通大学出版社
	（四川省成都市金牛区二环路北一段 111 号
	西南交通大学创新大厦 21 楼）
营销部电话	028-87600564　028-87600533
邮 政 编 码	610031
网　　　　址	http://www.xnjdcbs.com
印　　　　刷	四川煤田地质制图印务有限责任公司
成 品 尺 寸	170 mm × 230 mm
印　　　　张	8.5
插　　　　页	5
字　　　　数	139 千
版　　　　次	2024 年 8 月第 1 版
印　　　　次	2024 年 8 月第 1 次
书　　　　号	ISBN 978-7-5643-9945-0
定　　　　价	45.00 元

图书如有印装质量问题　本社负责退换
版权所有　盗版必究　举报电话：028-87600562

PREFACE 前言

磷（P）是引起水体富营养化的关键元素之一，复杂的污染来源排放，导致流域水体磷的迅速富集和水质恶化，磷的源解析是控制和预防水域水体磷污染的重要前提。近年来，磷污染已成为影响流域地表水环境质量安全的一个关键制约因素，有效治理地表水体磷污染的首要问题是准确找到磷污染源头并有效揭示其转化规律。随着社会经济迅速发展和人口不断增加，工业废水、生活污水和农田径流等排水影响，流域河流水体已逐渐演化成沿河两岸工农业生产和生活的纳污场所。近年来，主要水污染指标从化学需氧量（chemical oxygen demand，COD）和总氮（total nitrogen，TN）向总磷（total phosphorus，TP）转变，磷已成为河流水质超标的最重要因子之一。

以长江为例，长江是中华民族的母亲河之一，长江流域拥有约占全国20%的湿地面积、35%的水资源总量和40%的淡水鱼类种类，覆盖204个国家级水产种质资源保护区，是我国重要的生态安全屏障，亦是我国经济重心所在、活力所在，经济社会地位和生态环境价值突出。习近平总书记从中华民族永续发展的战略高度，亲自谋划、部署和推动实施长江经济带发展战略。近年来，长江生态环境保护已初见成效，但水资源、水环境、水生态、水风险等多重问题纷繁复杂、相互交织，水生态环境安全形势依然严峻。

基于第二次全国污染源普查数据，长江经济带11省市（包括上海市、江苏省、浙江省、安徽省、江西省、湖北省、湖南省、重庆市、四川省、云南省和贵州省）TP排放量分别来自农业源（68%）、生活源（30%）和工业源（2%）。

其中，农业源为区域污染物的主要来源，但农业源多为面源污染，其排放路径和入河量尚未清晰。尽管工业源仅占2%，但其入河系数较高，相比农业源对局部水体影响更直接，诸如云南玉溪地区、四川岷沱江流域为磷肥、黄磷、磷矿企业的集中区，江苏镇江—南通沿线为含磷农药和磷肥企业聚集区，这些工业源对流域的影响同样不可忽视。

2021年12月，国家发改委会同有关部门组织编制了《"十四五"重点流域水环境综合治理规划》（以下简称《规划》）。《规划》提出要深入打好污染防治攻坚战，加强大江大河和重要湖泊湿地生态保护治理，切实改善水环境质量。《规划》旨在进一步激发地方政府落实重点流域保护治理责任的积极性、主动性，发挥中央预算内投资的引导作用，吸引社会资本积极参与相关工程项目建设，推进重要湖泊和大江大河综合治理，助力深入打好污染防治攻坚战。

在此背景下，本书聚焦于日益凸显的流域磷污染，基于作者关于流域磷源解析的研究成果以及对当前研究现状的总结，系统地对流域磷来源、迁移转化过程以及磷源解析模型等进行介绍与讨论。全书共包括7章，第1章介绍了磷污染和磷的源解析概况，并基于文献计量学构建了全书的研究框架；第2章和第3章着重介绍了流域磷的来源、形态以及常用的表征手段，并基于这些内容对流域磷的迁移转化过程进行了总结；第4~6章分别从清单分析、扩散模型和受体模型三个方面对常用的磷源解析模型进行了详细介绍和分析，并在第7章对这些模型进行了综合讨论。

本书富有专业学科特色，各部分内容避免重复又相互呼应，力求构成完整的学术框架，反映学科前沿内容和国内外先进水平。本书有关的科研工作得到了国家自然科学基金（No. 51979237、No. 52170104、No. 52370102）和四川省自然科学基金面上项目（No. 23NSFSC0838）的资助，在此表示感谢。

　　本书内容涉及环境科学与工程专业知识，由于作者水平有限，书中难免有不足之处。我们期待本书在使用过程中能够得到专家学者和广大读者的批评指正，共同推动流域水环境安全和流域磷污染防控事业的可持续发展。

作　者

2024 年 8 月于成都

CONTENTS 目 录

第 1 章 绪　论 ·· 001
　1.1　流域磷污染概况 ··· 002
　1.2　文献计量学分析 ··· 003
　1.3　磷污染源解析研究框架 ··· 006

第 2 章 **流域磷来源、形态与表征** ··· 009
　2.1　磷来源 ··· 010
　2.2　磷形态 ··· 013
　2.3　磷的表征 ·· 015

第 3 章 **流域磷的迁移转化** ·· 031
　3.1　磷源的组成特征 ··· 032
　3.2　磷循环 ··· 033
　3.3　磷形态对流域的影响 ·· 038

第 4 章 **磷的源解析——清单分析法** ······································ 045
　4.1　清单分析的建立 ··· 046
　4.2　区域磷负荷清单分析 ·· 047
　4.3　磷负荷物质流清单分析 ··· 048
　4.4　行业生产清单分析 ·· 049
　4.5　清单分析综合评价 ·· 050

第 5 章　磷的源解析——扩散模型法 057
　　5.1　SWAT 模型 058
　　5.2　SPARROW 模型 065
　　5.3　AGNPS/AnnAGNPS 模型 069
　　5.4　SWMM 模型 073
　　5.5　扩散模型综合评价 077

第 6 章　磷的源解析——受体模型法 091
　　6.1　多元统计分析 092
　　6.2　稳定同位素法 100

第 7 章　磷的源解析方法综合评述 115
　　7.1　模型特性定量分析 116
　　7.2　源解析方法评分 117
　　7.3　关于源解析方法与磷形态的讨论 117

附　录 123
　　附录一　术语表（按首字母排序） 124
　　附录二　本书彩色插图 128

第1章

绪 论

1.1 流域磷污染概况

1.1.1 磷污染现状

磷（P）是一种具有重要生态功能的元素，也是生命中必不可少的元素。同时，磷在农业、工业、医药和能源领域均有着重要的用途。然而，过量的磷进入水体会引起严重的富营养化问题。目前，磷已经成为全球增长最迅速的污染物之一。1950—2000 年间，全球土壤中氮和磷负荷分别翻了 7.67 倍和 44 倍（Bouwman et al., 2013）。在中国，1978—2017 年间，农业面源污染中化学需氧量（Chemical Oxygen Demand，COD）、总氮（Total Nitrogen，TN）和总磷（Total Phosphorus，TP）负荷分别增长了 91.0%、96.2%和 244.1%（Zou et al., 2020）。磷污染已经成为制约流域生态环境可持续发展的关键因素。

磷在地壳中的丰度约为 0.12%，几乎都以矿物形式（如磷灰石，$Ca_5(PO_4)_3[F, OH or Cl]$）存在（Choi et al., 2009）。与氮不同，由于磷在自然界中无法形成完整闭合的循环，是一种不可再生资源（Keller et al., 2023）。然而，养分流失造成了严重的磷资源浪费。在中国，四川、湖南和湖北等农业大省化肥有效利用率不足 1/3，畜禽废弃物有效处理率不足 1/2（Huang et al., 2021; Zou et al., 2020）。在农业发达的美国，农作物对化肥和粪肥的利用率也不足 50%（Sabo et al., 2021）。Wang et al.（2022）发现肥料中 80%~90%的磷可能流失到了环境中。

1.1.2 磷污染源解析

磷污染已成为影响流域水环境质量安全的一个重要制约因素，有效治理河流水体磷污染的首要问题是准确找到磷污染源头并揭示其转化规律。流域水环境污染源解析方法包括：清单分析法、扩散模型法和受体模型法。清单分析法通过对污染源排放特征和地理分布等因素的观测和模拟，构建污染源信息列表自上而下地核算各污染源排放量或入河量；扩散模型通过模拟从源到汇的水文过程和污染迁移转化过程实现水体污染源解析，主要模型有 SWAT、SPARROW、AGNPS 和 SWMM 等；受体模型法是通过测量流域内不同位置受体的理化性质，定性地识别受体中的污染源，定量地计算污染源对受体的贡献率。在磷污染源解析方面，应用较广的受体模型有多元统计模型和稳定同位素。

清单分析法需要详细的污染源数据，计算过程复杂，工作量大，并且排放参数对解析结果影响大，具有较大不确定性，对河流污染防治实际指导效果有限；扩散模型多建立在物理机制基础上，对数据量和数据精度要求高，对于基础资料数据匮乏的河流，该法的应用受到严重限制；受体模型法无法从机理上表达污染物的生物化学转化过程，从而使识别结果带有一定主观性。以上关于河流磷污染的研究方法，从总体上可以估算磷来源及流通状况，但对污染来源的准确量化与磷的环境迁移转化认识还有不足，对于引起变化的内、外作用机制也待深入系统研究。由此，本书将结合流域中的磷污染来源与磷形态，深入分析常用的流域磷源解析模型的特点。为了建立完整的研究框架，首先对该领域涉及的研究进行文献计量学分析。

1.2 文献计量学分析

1.2.1 出版物数量分析

基于 Web of Science（WOS）数据库，对磷来源和溯源模型相关出版物数量进行了统计，见图 1-1。

如图 1-1（a）和（b）所示，磷来源相关出版物数量以较快速度逐年增长，到 2022 年达到 2959 项。其中，流域磷占比较大，面源磷相关出版物数量同样逐年增长。如图 1-1（c）所示，检索了磷污染溯源常用的几种方法（如清单分析、扩散模型和受体模型），可以看出扩散模型法（如 SWAT、SPARROW、AGNPS/AnnAGNPS 和 SWMM）和受体模型法（如多元统计法和稳定同位素法）出版物数量占比较大，其中 SWAT 是最常用的磷污染溯源模型，常年的研究占比高于 40%。

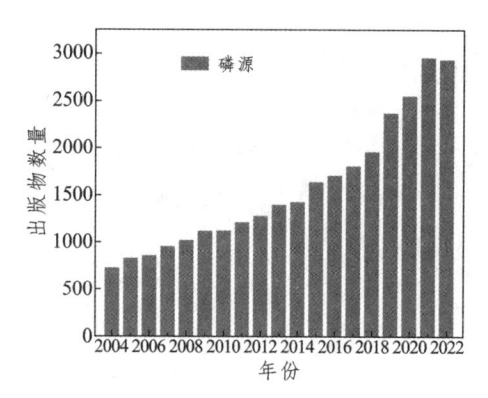

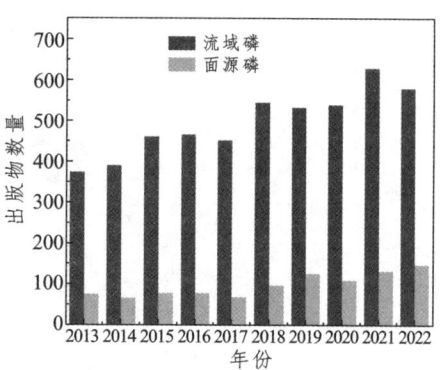

(a)

(b)

(c)

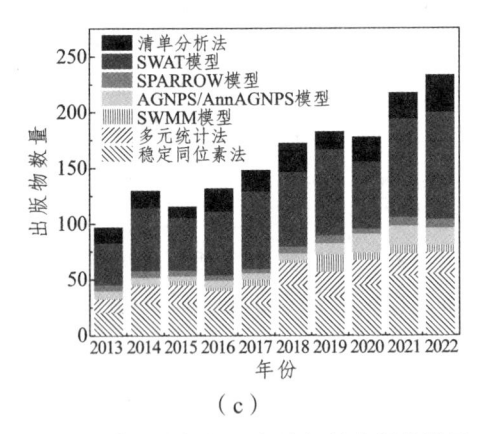

图 1-1　磷污染与溯源方法相关出版物数量

1.2.2　出版物聚类分析

基于 WOS 数据库，根据"磷来源""流域磷""面源磷""清单分析""扩散模型"和"受体模型"等关键词检索了数以万计的文献。筛选了与本研究相关性较强的 1535 篇文献，利用 VOSviewer 工具对这些文献进行了聚类分析，文献所属国家共线性网络见图 1-2，基于关键词的共线性网络见图 1-3。

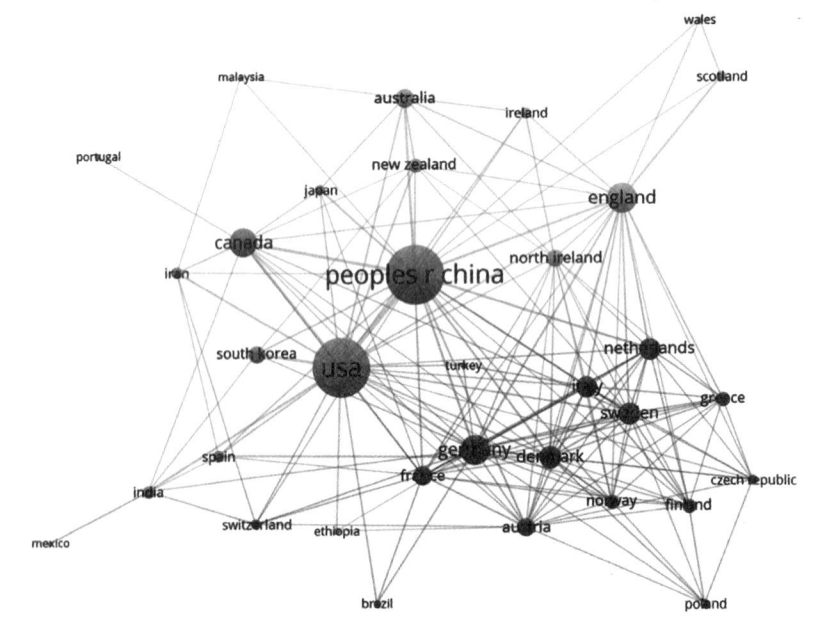

图 1-2　基于文献所属国家的文献聚类分析

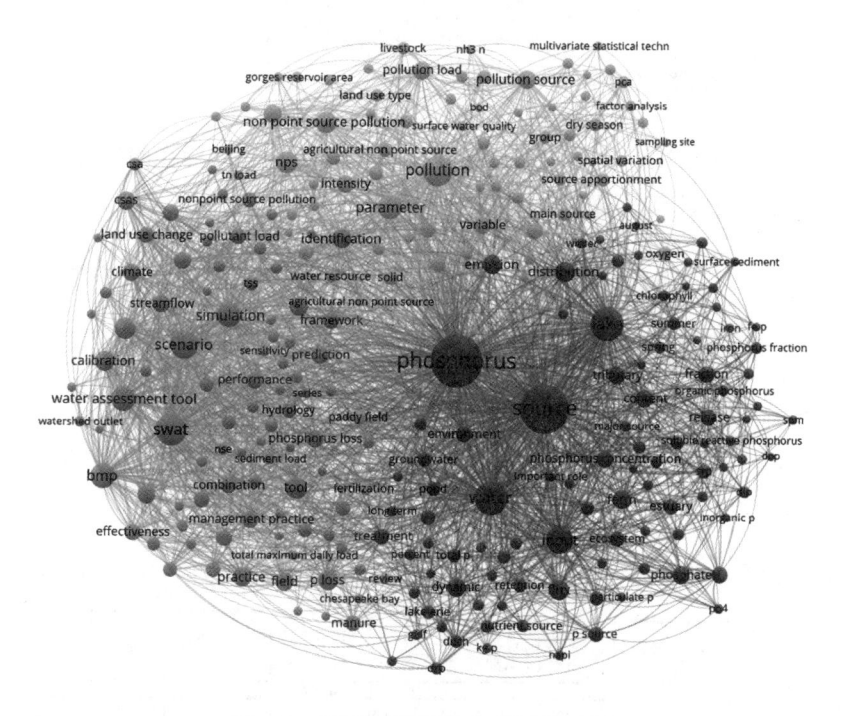

图 1-3　基于关键词的文献聚类分析

如图 1-2 所示（彩图见附录 2），开展磷污染和污染溯源研究的国家主要集中在中国、美国、加拿大和英国等 30 多个国家，可见磷污染已经成为一个全球化的研究热点。

如图 1-3 所示（彩图见附录 2），基于摘要对 1535 篇文献进行聚类分析，共筛选出现 5 次以上关键词 552 个，这些关键词可以分为三类。图中黄色部分，以磷污染为代表，包括了面源污染、污染负荷和源解析等；红色部分，以磷源为代表，还包括正磷酸盐（PO_4^{3-}）、溶解性有机磷（Dissolved Organic Phosphorus，DOP）、溶解性无机磷（Dissolved Inorganic Phosphorus，DIP）和颗粒磷（Particulate Phosphorus，PP）等磷形态；青色部分，以 SWAT 等溯源模型为代表，还包括了溯源模型的应用途径[比如，关键源区识别（Critical Source Areas，CSA）和最佳管理措施（Best Management Practice，BMP）等]。关键词共线性网络分析结果表明，以磷为中心的磷污染和污染源解析是一个十分庞大的研究体系，磷与流域水体联系十分紧密，其污染的精准溯源是一个亟待解决的关键难题。

1.3 磷污染源解析研究框架

基于上一节对磷污染与溯源相关出版物数量分析和共线性网络聚类分析，总结了该领域的研究框架，详见图1-4。

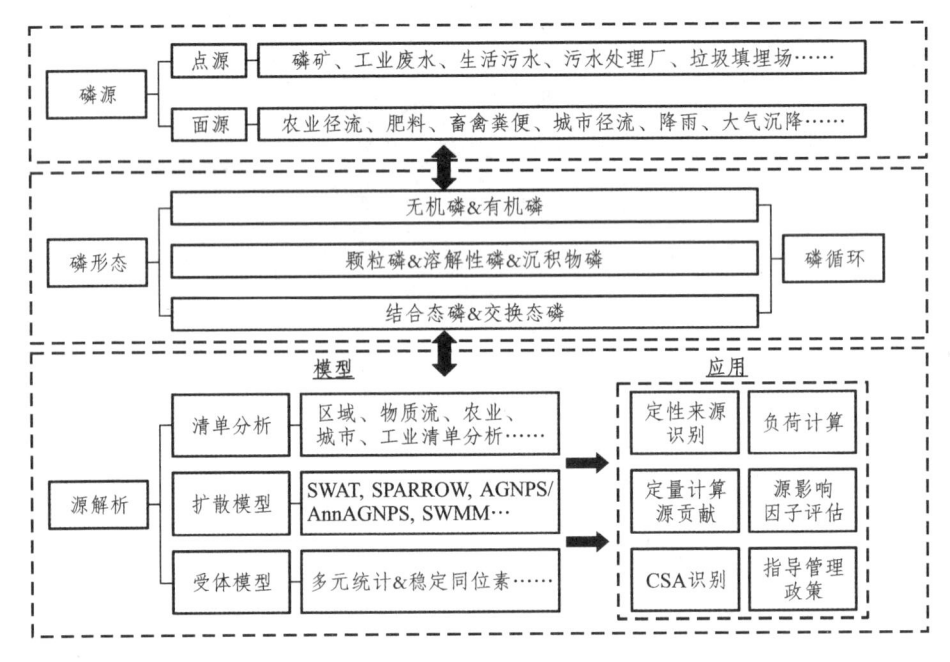

图1-4 磷污染溯源研究框架

如图1-4所示，磷污染溯源相关研究可分为磷污染、磷形态和源解析三个部分，这正好与图1-3中的文献共线性网络图的聚类结果相对应。本书将根据上述研究框架，分别对流域磷的来源、磷的形态与表征、磷的迁移转化以及磷的源解析模型等方面进行介绍。

参考文献

[1] Bouwman L, Goldewijk K K, Van Der Hoek K W, et al. 2013. Exploring global changes in nitrogen and phosphorus cycles in agriculture induced by livestock production over the 1900—2050 period[J]. Proc Natl Acad Sci, 110(52): 20882-20887.

[2] Choi H J, Choi C H, Lee S M. 2009. Analyses of phosphorus in sewage by fraction method[J]. J. Hazard. Mater. 167(1-3), 345-350.

[3] Huang J, Zhang Y, Bing H, et al. 2021. Characterizing the river water quality in China: Recent progress and on-going challenges[J]. Water Res. 201, 117309.

[4] Keller A A, Garner K, Rao N, et al. 2023. Hydrological models for climate-based assessments at the watershed scale: A critical review of existing hydrologic and water quality models[J]. Sci. Total Environ. 867, 161209.

[5] Sabo R D, Clark C M, Gibbs D A, et al. 2021. Phosphorus Inventory for the Conterminous United States (2002-2012)[J]. J. Geophys. Res. Biogeosci. 126(4), 1-21.

[6] Wang C, Luo D, Zhang X, et al. 2022. Biochar-based slow-release of fertilizers for sustainable agriculture: A mini review. Environ. Sci. Ecotechnol. 10, 100167.

[7] Zou L, Liu Y, Wang Y, et al. 2020. Assessment and analysis of agricultural non-point source pollution loads in China: 1978-2017[J]. J. Environ. Manage. 263, 110400.

第 2 章

流域磷来源、形态与表征

流域水体正面临着各类复杂来源磷的影响，从而带来了严重的富营养化风险。这些来自不同污染源的磷以溶解态、颗粒态、交换态和结合态等多种形式赋存在水体、悬浮物和沉积物等多种介质中。由于不同来源，不同形态磷的活性和生物有效性存在差异，其对流域环境的潜在影响也不尽相同。对不同形态的磷进行表征，分析其化学组成、组分和生物利用度有助于了解不同形态的磷对流域富营养化的具体影响。因此，本章总结了流域主要的磷源，梳理了流域中磷的主要赋存形态，介绍了流域中磷的表征手段，为后续章节中分析流域磷的迁移转化以及磷的源解析做铺垫。

2.1 磷来源

通常情况下，由流域生态系统外部输入的磷，其来源可以分为点源和面源（Zhao et al., 2023）。点源污染是指有固定排放点的污染源，多为工业废水及城市生活污水，由排放口集中汇入江河湖泊等水体（Du et al., 2022a; Xue et al., 2022）。面源污染是相对点源污染而言，指溶解态和颗粒态的磷从非特定的地点，在降水冲刷作用下，通过径流过程而汇入受纳水体（包括河流、湖泊、水库和海湾等）的过程（Ongley et al., 2010; Xue et al., 2022）。

2.1.1 点源

磷的点源污染通常由管道、通道或其他明确的物理结构，从特定的位置或设施中释放。主要的来源包括工业废水、污水处理厂（Wastewater Treatment Plant, WTTP）尾水、磷矿废水、垃圾渗滤液等（图 2-1, Oldfield et al., 2020）。相对面源来说，其排放的位置、排放量和排放成分是确定的，因此更容易追踪和控制（Xue et al., 2022）。不过，对于磷的点源污染目前仍存在一些争议。例如，大规模集约化的畜禽养殖所产生的磷排放通常认为是点源，然而在中国农村地区，大量的畜禽处于散养状态，这些畜禽产生的含磷废物可能会被归为面源污染（Ongley et al., 2010）；另外，化粪池系统常被认为是潜在的重要养分来源，但化粪池系统作为点源对流域养分负荷的贡献很少被量化（Oldfield et al., 2020）。

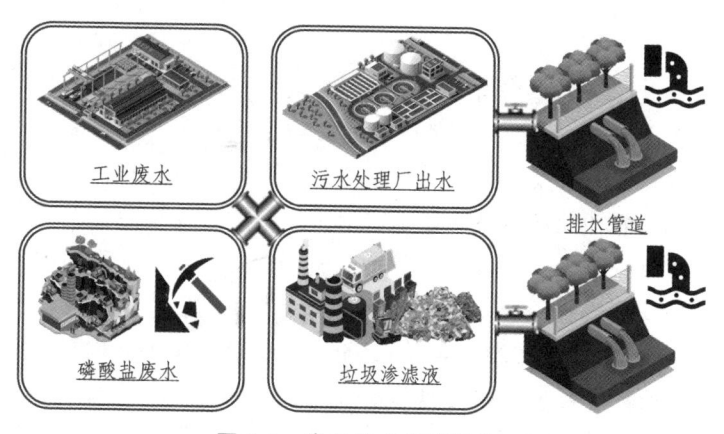

图 2-1 常见的点源磷污染

2.1.2 面源

面源污染是指污染物从非特定的地点，在降水冲刷作用下，通过径流过程而汇入受纳水体（包括河流、湖泊、水库和海湾等）的过程（Huang et al., 2021; Xie et al., 2022）。磷的面源污染主要来自农业面源、城市面源和大气沉降三个方面（图 2-2，Ongley et al., 2010）。与点源相比，面源污染具有随机性、分散性和隐蔽性，在时空上具有异质性（Shao et al., 2020; Xue et al., 2022）。目前，点源污染在许多国家已经得到较好的控制和治理，而面源污染，由于涉及范围广、控制难度大，已成为影响水体环境质量的重要污染源。Li et al.（2022a）研究表明，2017 年长江上游点源 TP 负荷为 0.04 Mt，而面源 TP 负荷达到了 4.64 Mt。

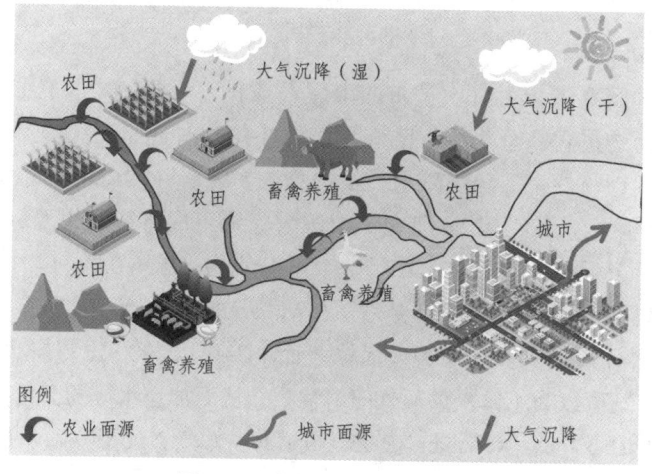

图 2-2 常见的面源磷污染

1. 农业面源

磷的农业面源污染是指由于农业灌溉、农业径流和土壤侵蚀等因素，导致磷从农田流失进入流域水体的现象（Ongley et al., 2010）。农业面源具有时空变化过程和机制复杂的特点，具有任意性和不规则性（Zou et al., 2020）。目前，在全球范围内，农业面源污染均是亟待解决的问题。在欧洲，农业对水生系统总污染负荷贡献约为55%；在农业发达的美国，农作物对化肥和粪肥的利用率也不足50%，这使得美国67%的湖泊和水库，53%的溪流暴露在农业面源污染当中（Chang and Zhang, 2023; Niraula et al., 2013）；在中国，化肥农药利用率低于三分之一，畜禽废弃物有效处理率低于50%（Huang et al., 2017）。农业生产依赖于生产要素的集约投入，这种农业模式对水环境造成了严重的面源污染（Zhang et al., 2019）。

2. 城市面源

磷的城市面源污染是指在晴天时累积在城市地表的磷，在暴雨的冲刷下，通过地表径流或管道进入受体水域，造成水污染的现象（Huang et al., 2022）。快速城市化带来的土地利用和土地覆盖的显著变化，造成不透水面积激增，地表水文发生强烈变化，减少了入渗，增大了径流量，并最终导致土壤侵蚀和养分流失（Li et al., 2022c; Shao et al., 2020; Shrestha et al., 2018）。在中国，北京和上海中心城区，城市面源污染对水污染的贡献率高达50%（付超 等, 2020）。然而，土地利用，地表入渗，降雨强度和干旱天数的变化均会造成城市面源污染的高异质性，导致污染来源的追踪和控制难度增大（Huang et al., 2022）。

3. 大气沉降

磷的大气沉降是指大气中携带的磷颗粒或化合物通过空气传播并沉降到地面或水体表面的过程，分为干沉降和湿沉降（Huang et al., 2017）。大气中磷的主要来源包括燃料燃烧、矿物粉尘和植物释放（Li et al., 2020; Li et al., 2021; Ma et al., 2022）。Ma et al.（2022）研究了中国大气磷组成特征及来源，表明在2019—2021年间，中国大气中PP、DIP和DOP分别占TP的24%、43%和33%。其中，人为的燃烧是大气中磷的主要来源（Ma et al., 2022）。这些磷通过气溶胶、颗粒物和降水等形式沉积到地表或水体表面（Chiwa, 2020）。尽管大气中的磷浓度相对较低，却可以为水体提供额外的磷输入，从而加速水体的富营养化（Chiwa, 2020; Okin et al., 2004）。Newman（1995）报道称，大气P沉积[0.07~1.7 kg/(ha·a)]与陆地生态系统中岩石风化产生的P[0.05~1.0 kg/(ha·a)]相当，表明大气沉降是磷面源污染的不可忽略的途径。植物释放是指在生长季节，植物将磷以磷酸三乙酯形式向大气中释放，并迅速转移到颗粒物上（Li et al., 2021）。

2.2 磷形态

在流域中，磷有多种赋存形式，存在于水体、悬浮颗粒物和沉积物等多种介质中（Ni et al., 2022）。在水体和悬浮颗粒物中，按化学形态分类，可分为无机磷（P_i）和有机磷（P_o）；按物理形态分类，可分为颗粒态磷（Particulate Phosphorus，PP）和溶解态磷（Dissolved Phosphorus，DP）（Bai et al., 2009; Du et al., 2022b）。由此形成了溶解性无机磷（Dissolved Inorganic Phosphorus，DIP）、溶解性有机磷（Dissolved Organic Phosphorus，DOP）、颗粒态无机磷（Particulate Inorganic Phosphorus，PIP）和颗粒态有机磷（Particulate Organic Phosphorus，POP）等形态。在沉积物中，根据 P 与沉积物等介质的结合紧密程度分为交换态磷（Exchange Phosphorus，Ex-P）和结合态磷（Bound Phosphorus）。本节按化学形态类别罗列了 P_i 和 P_o 在水体、悬浮颗粒物和沉积物中的不同存在形式，并系统分析了这些磷形态的特性。进一步地，详细介绍了相对活跃的生物有效磷（Bioavailable Phosphorus，BAP），为后文讨论磷的活性与生物有效性做铺垫。

2.2.1 无机磷

无机磷（P_i）是指流域中以无机化合物形式存在的磷，这部分 P_i 常以磷酸盐的形式存在，包括：正磷酸盐（ortho-P，PO_4^{3-}）、偏磷酸盐（meta-P，PO_3^-）、焦磷酸盐（pyro-P，$P_2O_7^{2-}$）和聚磷酸盐（Polyphosphates）等（Aydin et al., 2009）。P_i 极容易被水体中的藻类和浮游生物吸收，是导致水体富营养化的主要原因之一（Pu et al., 2023）。其中，正磷酸盐是所有磷形态中生物利用度最高的形式（Feng et al., 2020）。

按照赋存的物理形式，P_i 分为溶解态 P_i、颗粒态 P_i、结合态 P_i 以及交换态 P_i。其中水体中主要的赋存形式为溶解态 P_i、颗粒态 P_i。溶解性 P_i 的生物利用度极高，是溶解性活性磷（Soluble Reactive Phosphate，SRP）的重要组成部分（Huser and Pilgrim, 2014）。颗粒态 P_i 直径通常在几微米到几十微米之间，可以悬浮在水体中。其生成和沉积受多种环境因素影响（如溶解氧、pH 值、水温、微生物等）（Hafuka et al., 2021）。

结合态 P_i 和交换态 P_i 是沉积物中磷的主要赋存形式，通常通过分级提取法

（Aydin et al., 2009）所得，按照提取顺序依次为铝结合磷（Al-P）、铁结合磷（Fe-P）、钙结合磷（Ca-P）、有机磷（P_o）和残余磷（Res-P）等（Cao et al., 2016; Choi et al., 2009; Liu et al., 2015; Peng et al., 2019; Zhu et al., 2023）。详细提取过程将在 2.2.2 节中阐述。其中，Fe-P 和 P_o 活性较高，是沉积物中相对活跃的一种磷。而 Ca-P 主要以磷灰石的形式存在，在沉积物中相对稳定（Bai et al., 2009; Du et al., 2022b; Zhu et al., 2023）。分级提取后的 Res-P 是更稳定的磷形态（Liu et al., 2015）。

交换态 P_i（Ex-P）是相对于结合态 P_i 一种存在形式。这类磷通常松散吸附在沉积物，黏土矿物和悬浮颗粒表面。Ex-P 会与 OH^-、Br^-、Cl^- 和 SO_4^{2-} 竞争吸附点位，因此很容易进入水体，被初级生产者直接吸收（Chen et al., 2019; Saha et al., 2023）。

2.2.2 有机磷

有机磷（P_o）在流域中组成较为复杂，其包括两个重要的键[如磷脂（C-O-P）和磷酸（C-P）]，其中 C-O-P 可被碱性磷酸酶（Alkaline Phosphatase，APA）降解，释放 P_i，而磷酸（C-P）相对磷脂难降解（Jia et al., 2022; Jónasdóttir, 2019）。按照物理形式可分为溶解态 P_o（Dissolved Organic Phosphorus，DOP），颗粒态 P_o（Particulate Organic Phosphorus，POP）（Yang et al., 2019）。

与 P_i 相似，P_o 可以通过分级提取法获得，包括：活性 P_o（labile P_o，LP_o）、中等活性 P_o（Moderately Labile P_o，MLP_o）和非活性 P_o（Nonlabile P_o，NLP_o）等存在形式。其中，LP_o 包括单酯磷（Mono-P）、二酯磷（Diester-P）等，被认为是生物利用性与 P_i 相当的 P_o，由 $NaHCO_3$ 提取，亦可称为 $NaHCO_3$-P_o（Zhu et al., 2023）；MLP_o 包括由 HCl 提取的 HCl-P_o 和 fulvic-P_o（Ful-P_o），HCl-P_o 是一类由 P_o 与植酸、Ca 和 Mg 结合形成的磷，Ful-P_o 则是 P_o 与富里酸结合形成的磷；NLP_o 包括 Humic acid-P_o（Hum-P_o）和 Res-P_o，Hum-P_o 包括五磷酸肌醇、六磷酸肌醇等难降解物质组成，而 Res-P_o 由 P_o 与植酸结合组成（Ni et al., 2020; Pu et al., 2020; Tao et al., 2022; Zhu et al., 2023）。

除此之外，由于有机磷阻燃剂（Organo-Phosphorous Flame Retardant，OPFRs）和有机磷农药的广泛使用，流域水体中检测出包括三苯基磷酸（TPhP）和敌敌畏（DDVP）等一系列具有毒性、致癌性和持久性的 OPFRs 和农药，这类 P_o 对于流域的影响同样不可忽视（van der Veen and de Boer, 2012; Wang et al., 2021）。

2.2.3 生物有效磷

为阐明流域中不同磷形态在磷循环中扮演的角色,将生物有效磷这一类磷单独进行讨论。BAP 是指可被水体及沉积物中的生物直接利用的磷,其代表的是流域水体和沉积物中最活跃的一类磷,同样也是磷循环中最核心的存在形式(Pu et al., 2023)。明确了 BAP 的组成,可以进一步厘清其在流域磷循环中的转化途径与机理。在水体中所有的溶解性 P_i 均为 BAP。此外,部分溶解态 P_o 和颗粒态 P_o 也属于这一范畴。溶解态 P_o 具有很高的潜在生物利用度和通过生物矿化和光矿化补偿 P_i 的能力,这部分 P_o 以 Mono-P 和 Diester-P 为主(Li et al., 2022b)。

在沉积物中,BAP 被认为是 Ex-P、Fe-P 和 P_o 的总和。在这之中,由于 Ex-P 通常松散吸附在沉积物中,极易被生物直接利用;Fe-P 是结合态 P_i 中最活跃的组分,在环境因素变化时,尤其是还原环境下,很容易释放到上覆水中(Pu et al., 2020);沉积物中的 P_o 通常由核酸、核酸肌醇和磷脂组成,很容易在酶的作用下矿化,释放 P_i,也被认为是水生植物和浮游动物的潜在磷源(Pu et al., 2023)。

2.3 磷的表征

磷的表征是了解其性质、确定其环境行为、归趋和潜在环境风险的关键。从最基本的 TP 浓度、PP 和 DP 浓度等表征流域整体污染水平,到通过光谱学或质谱学方法直接指示磷的组成和结构,不同的表征方法是揭示磷在流域中扮演不同角色的有力手段。

2.3.1 磷浓度的测定

通常情况下,流域水体的磷污染水平是通过 TP、总溶解性磷(Total Dissolved Phosphorus,TDP)和总颗粒磷(Total Particulate Phosphorus,TPP)等浓度来反映的。TP 和 TDP 均可通过钼酸铵分光光度法(GB 11893—1989)测定(Pu et al., 2023)。具体地,将水样在中性条件下用过硫酸钾消解,使样品中所有形式的磷(DP、PP、P_i 和 P_o)全部氧化为 ortho-P。在酸性介质中,正磷酸盐与钼酸铵反应,在锑盐存在下生成磷钼杂多酸后,立即被抗坏血酸还原,生成蓝色络合物。显色后使用分光光度计在 700 nm 波长下测试其吸光度,从而得到 TP 的浓度。而 TDP 则是通过过滤或离心,将水样中的颗粒物去除,再通过钼酸铵分光光度法测定。TPP 则根据式(2-1)计算。

$$C_{\text{TPP}} = C_{\text{TP}} - C_{\text{TDP}} \qquad (2\text{-}1)$$

2.3.2 磷的分级提取

前文提到了 P_i 和 P_o 均可通过分级提取分馏出不同的组分，本节将着重阐述分级提取的方法，包括详细步骤、使用的萃取剂以及对应的组分。

1. P_i 的分级提取

P_i 的顺序提取法如图 2-3 所示，使用的试剂包括 1.0 M NH_4F、1.0 M NH_4Cl、0.5 M NH_4F、0.1 M $NaOH$ 和 1 M H_2SO_4，均为分析纯。根据不同的提取剂依次得到 Ex-P、Al-P、Fe-P、Ca-P、P_o 和 Res-P 等组分。其中 P_o 和 Res-P 会在后文中进一步阐述提取步骤，得到的组分可以通过电感耦合等离子体原子发射光谱仪（Inductively Coupled Plasma-Atomic Emission Spectrometer，ICP-AES）在 213.617 nm 波长下测定其浓度（Aydin et al., 2009; Choi et al., 2009）。

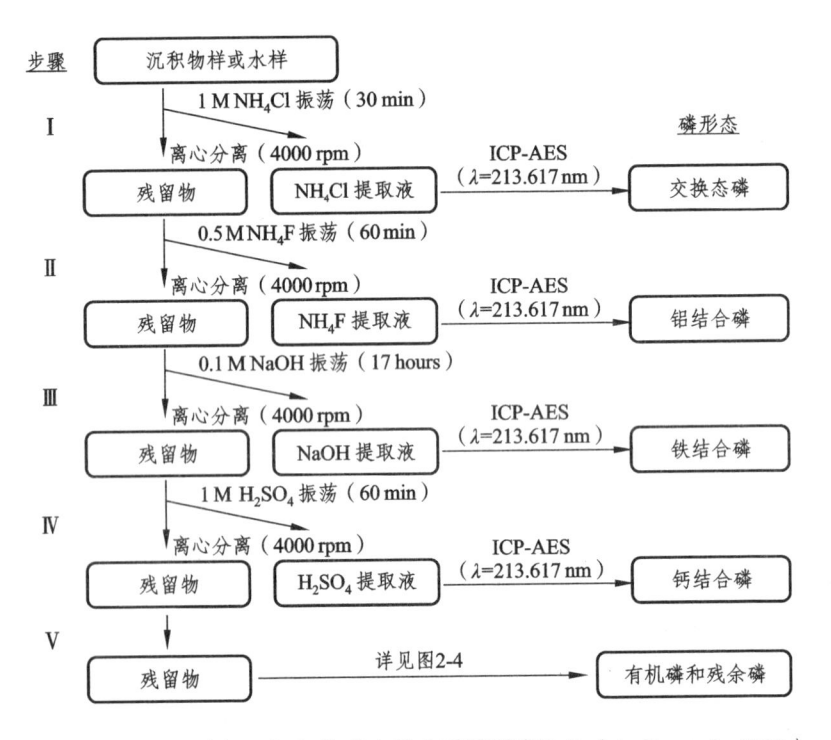

图 2-3 顺序提取法提取沉积物或水样中不同形态的 P_i（Aydin et al., 2009）

2. P_o 的分级提取

采用 Ivanoff 等人（1998）开发的顺序提取法，分有机磷五步工艺从沉积物中依次提取 P_o 组分，见图 2-4。使用萃取剂包括 0.5 M $NaHCO_3$、1M HCl、0.5 M NaOH 和 1 M H_2SO_4。依次获得组分为活性 P_o（labile P_o，LP_o）、中等活性 P_o（moderately labile P_o，MLP_o）和非活性 P_o（nonlabile P_o，NLP_o）。在每个提取步骤中，均使用钒钼酸盐法分光光度法测定提取组分中的 TP 和 P_i 含量，由此计算各组分 P_o 含量（Ni et al., 2020）。

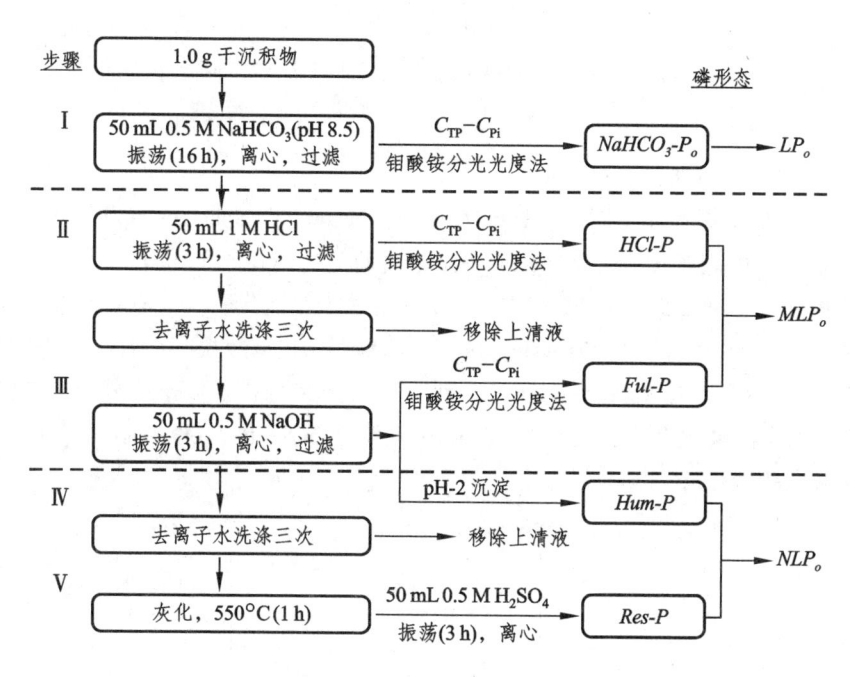

图 2-4 顺序提取法提取沉积物中不同形态的 P_o（Chen et al., 2019）

2.3.3 P_o 的组成与结构

相对于 P_i，无论是在水体还是沉积物中，P_o 的组成和结构尤为复杂，对流域富营养化的影响机制尚未得到完整清晰的认识。而 P_o 组成和结构的表征是研究沉积物 P_o 的关键。因此，本节将分别从光谱学和质谱学的角度，着重介绍 P_o 组成和结构的表征方法（图 2-5）。

```
化学组成          官能团           磷形态                      分子组成
(³¹P NMR)       (FT-IR)         (X射线荧光光谱,XANES)        (FT-ICR MS)
```

- 单酯磷
- 二酯磷
- 膦酸酯

- 磷脂, DNA-P, RNA-P, P═O键
- 芳香族化合物 (P—O—C键)
- 脂肪族化合物 (P—O—C键)

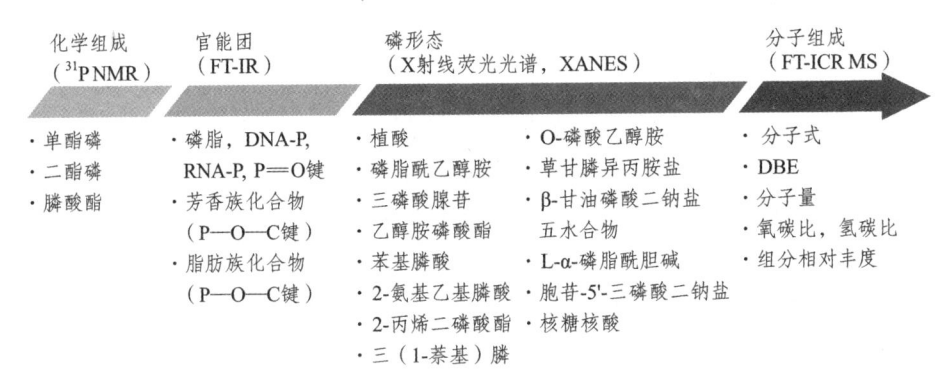

- 植酸
- 磷脂酰乙醇胺
- 三磷酸腺苷
- 乙醇胺磷酸酯
- 苯基膦酸
- 2-氨基乙基膦酸
- 2-丙烯二磷酸酯
- 三(1-萘基)膦

- O-磷酸乙醇胺
- 草甘膦异丙胺盐
- β-甘油磷酸二钠盐五水合物
- L-α-磷脂酰胆碱
- 胞苷-5'-三磷酸二钠盐
- 核糖核酸

- 分子式
- DBE
- 分子量
- 氧碳比,氢碳比
- 组分相对丰度

图 2-5 P_o 化合物组成和结构的常用表征方法(Ni et al., 2022)

1. 光谱学方法

(1)^{31}P 核磁共振光谱。

^{31}P 核磁共振光谱(^{31}P Nuclear Magnetic Resonance Spectroscopy,^{31}P NMR)是最常用的 P_o 表征手段。该方法具有非破坏性,可以识别 P_o 的化学组成,包括单酯 P(Monoester P, Mon-P,例如:α/β-甘油磷酸、核苷酸、D-葡萄糖 6-磷酸)、二酯 P(Diester P,例如:DNA-P、RNA-P 和磷脂)和磷酸盐等(Ni et al., 2022)。如图 2-6 所示,Hafuka et al.(2021)利用 ^{31}P NMR 分析了城市河流和下游富营养化湖泊中的磷化合物,从图中可以清晰识别出 Ortho-P、Monoester-P 和 DNA-P 等 P 形态的特征峰。

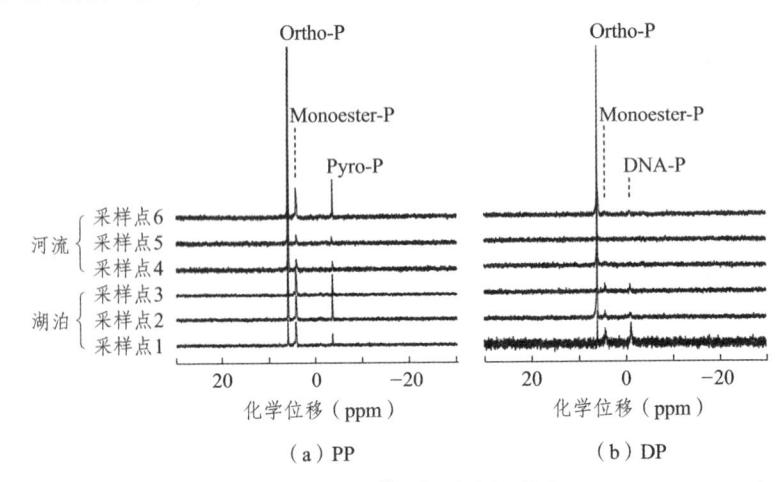

图 2-6 (a)PP 和(b)DP 溶液 ^{31}P 核磁共振谱(Hafuka et al., 2021)

值得注意的是,在 ^{31}P NMR 预处理过程中,乙二胺四乙酸酯(Ethylene Diamine

Tetraacet Ate，EDTA）-NaOH 萃取和富集过程会引起 P_o 化合物的化学降解，导致沉积物中 P_o 含量被低估（Ni et al., 2022）。

（2）傅里叶变换红外光谱。

傅里叶变换红外光谱（Fourier Transform Infrared Spectroscopy, FT-IR）是另一种非破坏性的方法，它提供了关于 P 化学键和溶解性有机物（Dissolved Organic Matters, DOMs）官能团振动的实时信息。如图 2-7 所示，1033～1030、876～873 和 797～779 cm^{-1} 处的 FT-IR 光谱主峰通常与磷脂、DNA 和 RNA-P（P=O 键）、芳香族 P 化合物（P－O－C 键）和脂肪族 P 化合物（P－O－C 键）有关（Ni et al., 2019）。不过，如果样品中含有结构相似的复杂化合物，它们的光谱峰将重叠，化合物将无法区分（Ni et al., 2022）。

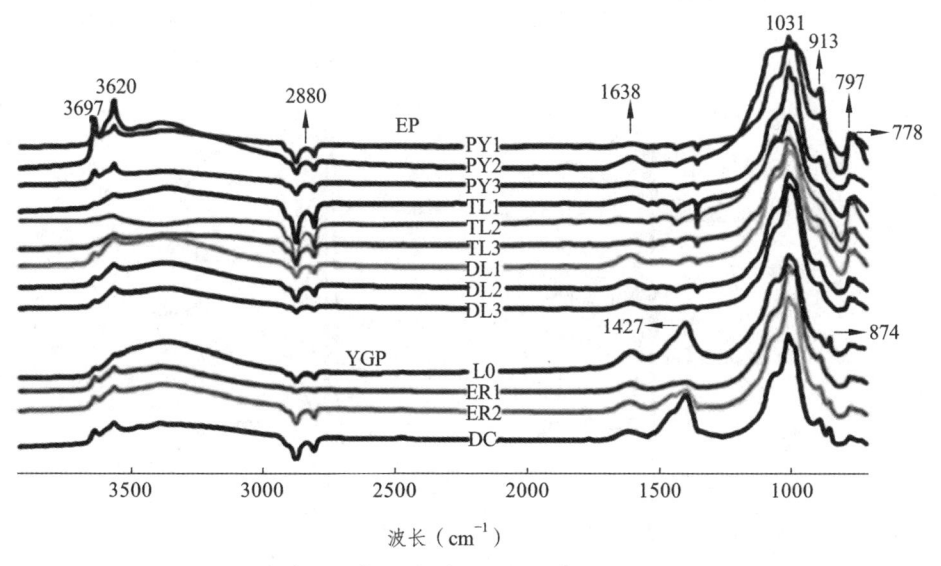

图 2-7　湖泊沉积物的典型 FT-IR 光谱（Ni et al., 2019）

（3）X 射线荧光光谱。

X 射线荧光光谱（X-ray fluorescence spectroscopy）和显微镜用于元素丰度分析和形态表征，灵敏度高达万亿分之一，可用于鉴定沉积物中的磷形态。如图 2-8 所示，Brandes et al.（2007）使用该方法鉴定了海洋沉积物中的磷形态，包括植酸（phytic acid）、磷脂酰乙醇胺（phosphatidylethanolamine）、5′-三磷酸腺苷（adenosine 5′-triphosphate）、邻-磷酸乙醇胺（O-phosphorylethanolamine）、苯基磷酸（phenyl phosphate）、2-氨基乙基膦酸（2-aminoethylphosphonic acid）、

苯膦酸（phenylphosphonic acid）、丙二膦酸（propylenediphosphonic acid）等。

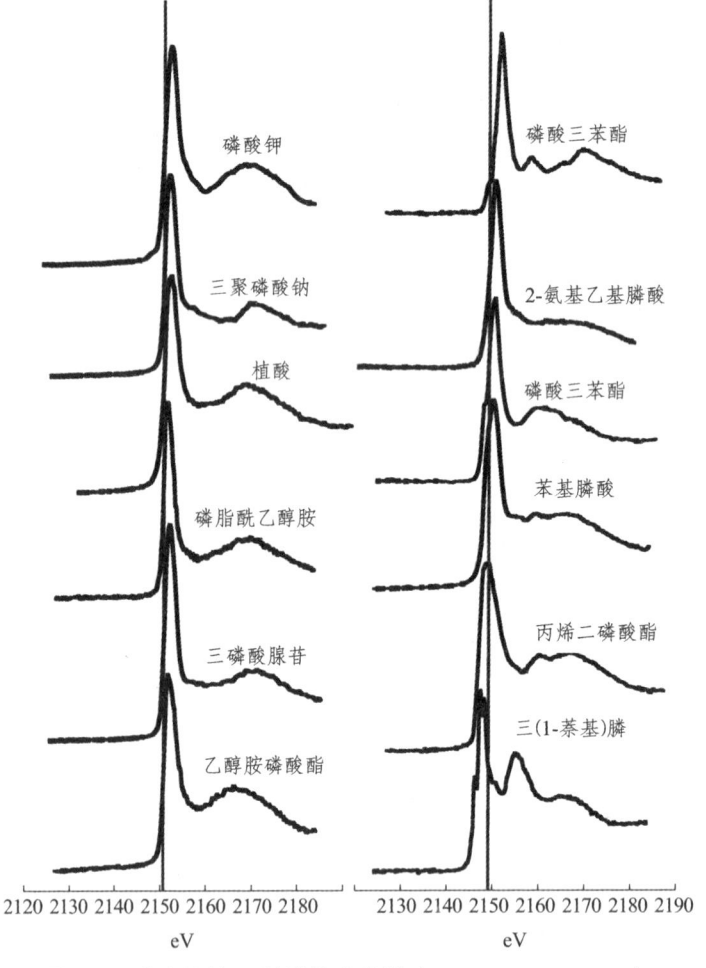

图 2-8 磷酸盐的 X 射线荧光光谱（Brandes et al., 2007）

（4）X 射线吸收近边结构光谱。

X 射线吸收近边结构光谱（X-ray Absorption Near-Edge Structure Spectroscopy, XANES）可以直接识别 P 形态，无需预提取和最小的样品预处理（Ni et al., 2022）。如图 2-9 所示，Kruse et al.（2009）利用 XANES 识别了包括邻-磷酸乙醇胺（O-phosphorylethanolamine）、2-氨基乙基膦酸（2-aminoethylphosphonic acid）、正（磷甲乙基）甘氨酸[N-（phosphonomethyl）glycine]、β-五水磷酸甘油二钠盐（β-Glycerol phosphate disodium salt pentahydrate）、1-α-磷脂酰胆碱

（l-α-Phosphatidyl choline）、1-α-磷脂酰乙醇胺（l-α-Phosphatidyl choline）、核糖核酸（ribonucleic acid）、植酸钠水合盐（ribonucleic acid）、植酸钙盐（ribonucleic acid）等。

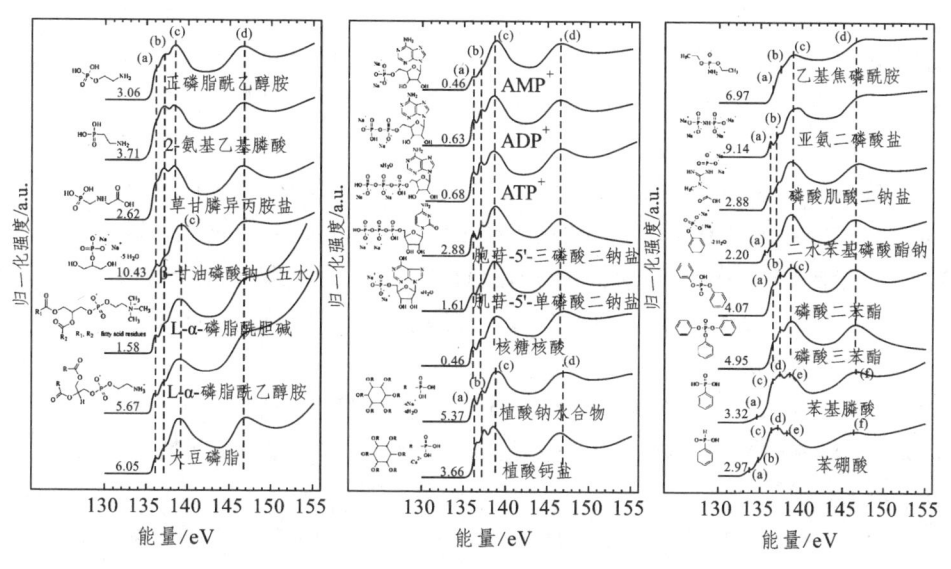

图 2-9　部分 P_o 的 X-ANES 光谱（Kruse et al., 2009）

2. 质谱学方法

P_o 的组成与结构表征主要的质谱学方法为傅里叶变换离子回旋共振质谱法（Fourier Transform Ion Cyclotron Resonance Mass Spectrometry, FT-ICR-MS）。该方法是一种较新的 DOM 表征手段，其允许根据分子式检测多达数千种的 DOM 成分。识别的参数包括双键当量（Double Bond Equivalent, DBE）、分子式、分子质量、O/C 和 H/C 比，以及 P_o（如缩合芳香结构、单宁、不饱和烃、木质素、脂质、蛋白质和碳水化合物）的相对丰度（Ni et al., 2022）。

在对 P_o 进行 FT-ICR-MS 分析时，共需三个步骤：① 预处理：通过对连续提取的 H_2O、$NaHCO_3$、NaOH 提取液经过固相萃取柱富集和提纯；② FT-ICR-MS 测试：将提取液进入在装有 9.4 t 超导磁体和阿波罗 II 电喷雾电离源的 FT-ICR-MS 中测试，采用 ESI 负离子模式以 120 μL/h 的流速连续进样，m/z 范围为 100~1600 Da；③ 分子式匹配：使用仪器自带的软件 Data analysis 5.2，根据 H 原子数和 C 原子数比值（H/C）≤$2n+2$（C_nH_{2n+2}），H/C 和 O/C 范围为 0.2<H/C<2.2 和 0.1<O/C<1.2，信噪比（S/N）>4，m/z<800 的原则剔除异常分子式，

匹配正确的分子式（Brooker et al., 2018; Kurek et al., 2021b; Pu et al., 2023）。

根据 FT-ICR-MS 获取的分子式，分析分子式的参数和组成特征。其中计算加权参数特征，包括 m/z_{wa}、C_{wa}、H_{wa}、O_{wa}、N_{wa}、P_{wa}、S_{wa}、H/C_{wa}、O/C_{wa}、识别有机磷的分子式参数差异。识别含有不同元素组合，如 CHON、CHONS、CHOP、CHONP、CHOSP、CHONSP 的分子组成差异，比较含 P 的分子化合物的特征。重点区分含 1 个和 2 个 N 原子，含 1 个和 2 个 P 原子，以及不包含 N 和 P 原子的分子参数特征（Pu et al., 2023; Zhou et al., 2022）。

Pu et al.（2023）使用 FT-ICR-MS 识别了洱海藻华高、低风险期沉积物中生物有效 P 的变化（图 2-10，彩图见附录 2）。根据 Van-Krevelen 图识别有机氮磷中分子化合物的相对丰度，判别分子化合物的类别为：① 脂质化合物（O/C = 0 ~ 0.3，H/C = 1.5 ~ 2.0）；② 蛋白质和氨基酸类化合物（O/C = 0.3 ~ 0.67，H/C = 1.5 ~ 2.2）；③ 氨基糖/碳水化合物（O/C = 0.67 ~ 1.2，H/C = 1.5 ~ 2.2）；④ 木质素类化合物（O/C = 0.1 ~ 0.67，H/C = 0.7 ~ 1.5）；⑤ 浓缩芳香类化合物（稠环芳烃）（O/C = 0 ~ 0.67，H/C = 0.2 ~ 0.7）；⑥ 单宁类化合物（O/C = 0.67 ~ 1.2，H/C = 0.5 ~ 1.5）。

此外 FT-ICR-MS 还可根据 DBE_{wa}、$AI_{mod,wa}$、$NOSC_{wa}$ 等参数，揭示分子稳定性特征，计算公式如下：

$$DBE_{wa} = \Sigma(C - 1/2H + 1/2N + 1)_i \times M_i \tag{2-2}$$

$$AI_{mod,wa} = \Sigma[(1 + C - 1/2O - S - 1/2H)/(C - 1/2O - N - S - P)]_i \times M_i \tag{2-3}$$

$$NOSX_{wa} = \Sigma\{[(1 + C - 1/2O - S - 1/2(N + P + H)]/(C - 1/2O - N - S - P)\}_i \times M_i \tag{2-4}$$

在式（2-2）中，DBE_{wa} 是代表有机氮磷分子化合物中双键和脂肪环（C=C 和 C=O）的总数，通过分子式中 C、H、N 原子数量计算而得。当 DBE 为 0 时，说明化合物完全饱和；DBE>0，说明随着双键/环数增加会伴随着 2 个 H 的损失，而化合物逐渐不饱和，DBE 越大表示物质不饱和程度越高（Chen et al., 2020a; Lusk and Toor, 2016）。

在式（2-3）中，$AI_{mod,wa}$ 包含 C、H、O、N、S 和 P 原子数量，该指数代表化合物的芳香程度，AI<0.5，且 H/C 值<1.5 表示来自于木质素降解的土壤不饱和物质；0.5<AI<0.66 表示酚类、醛类物质；AI>0.67 则反映物质主要为多环芳烃类物质（Chen et al., 2020b）。

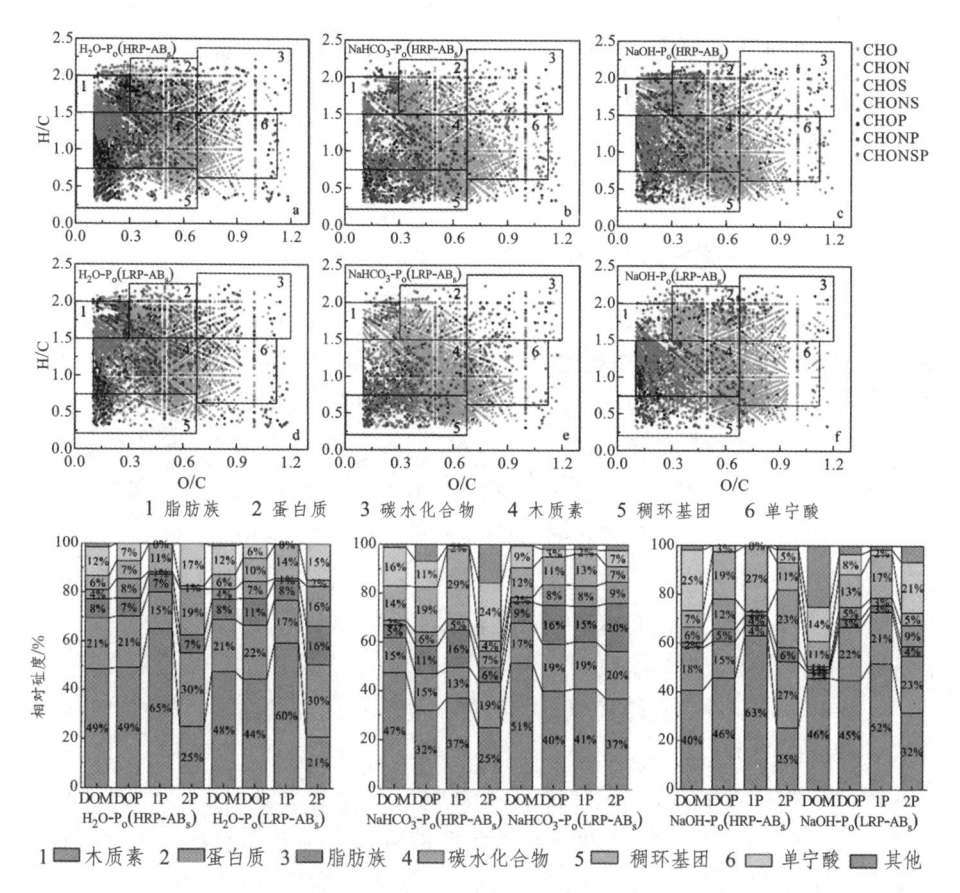

图 2-10　P_o 的分子组成的 Van-Krevelen 图以及相对丰度（Pu et al., 2023）

在式（2-4）中，$NOSC_{wa}$ 分别代表式中包含 C、H、O、N、S 和 P 原子，代表的是碳的标准氧化态，反映了有机质和有机磷分子的生物地球化学活性和生物有效性（Chen et al., 2020c）。

利用高分辨率质谱技术（FT-ICR-MS）为从分子角度探究复杂有机化合物提供了新的视角。可有效区分流域内畜禽养殖粪便、废水和河水中有机磷的分子组成差异。这反映了 FT-ICR-MS 技术揭示流域生态系统中有机磷分子组成受到来源、流域过程、生物和非生物相互作用等复杂影响而分异的特征（Brooker et al., 2018）。来自于陆源的植物残体、风化土壤等通常含有较高木质素类化合物，具有抗生物降解性（Zhou et al., 2022）；而湖内藻类、自生源动植物残体和微生物等贡献则以脂质、蛋白质和氨基糖类组成为主（Kellerman et al., 2015）。在湖泊遭受环境变化影响（藻华暴发前后），沉积物有机质表现出由脂肪族化合物组成为主向芳香类化合物组成为主的变化（Kurek et al., 2021a），指示湖泊沉积物有机化合物从分子组成上对生物地球化学过程的响应。沉积物有机氮磷的分子特征既能反映流域不同来源的贡献，也可以作为评估其潜在生物降解能力的依据，是探究生态系统氮磷生物地球化学过程的有效载体。

2.3.4 磷的生物有效性测定

磷的生物有效性是评价磷对湖泊富营养化潜在贡献的重要指标。生物有效性是指特定 P_o 被生物降解并被微生物生物量同化的相对容易程度（Duhamel et al., 2021）。它与微生物的摄取、同化和吸收能力以及 APA 活性密切相关（Ni et al., 2022）。P_o 生物有效性测定的常用方法包括酶水解和分子特性表征。

酶水解已被广泛用于估计 P_o 的生物有效性，这是由于大多数 P_o 在被生物体利用之前必须被水解成 SRP（Buenemann et al., 2008）。常用的酶主要来源于陆生生物，例如牛肠黏膜（碱性磷酸酶）、响尾蛇（磷酸二酯酶）、小麦（植酸酶）等。然而，Duhamel et al.（2021）研究表明使用酶水解进行生物有效性评估时，使用更多样化的水生生物酶更具代表性。酶水解法评估 P_o 的生物有效性易于应用，但由于环境条件的复杂性以及酶的种类和数量的复杂性，其精度和重现性较差（Ni et al., 2022）。

分子特性是水生系统中微生物利用 DOM 的主要控制因素（Shen and Benner, 2020）。了解 P_o 分子特性，有助于评价天然群落对不同 P_o 的反应。P_o 分子特性通常包括组成、结构、疏/亲水性、分子量、腐殖化程度、官能团数量、化学计量比等（Ni et al., 2022）。上述特性可以通过紫外-可见光谱（Ultraviolet and Visible Spectrophotometry, UV-Vis）、三维荧光光谱（Excitation-Emission-Matrix Spectra, EEM）和 FT-ICR-MS 等手段表征。例如，LP_o 大约 3 天内即可矿化，而其他 P_o 矿化则需要 4~13 周（Darch et al., 2014）。在 DOP 中 C/P 较低的疏水 DOP 组分对藻类生长的生物可利用性更强。Gao et al.（2021）研究表明低氧、低芳香性和低不饱和度的 DOP 在污水处理厂被生物处理优先降解。Ohno et al.（2014）发现，森林土壤中 H/C>为 1.2 和 O/C>为 0.5 的 DOM 组分与生物降解性密切相关。

参考文献

[1] Aydin I, Aydin F, Saydut, A, et al. 2009. A sequential extraction to determine the distribution of phosphorus in the seawater and marine surface sediment[J].

J. Hazard. Mater. 168(2-3), 664-669.

[2] Bai X, Ding S, Fan C, et al. 2009. Organic phosphorus species in surface sediments of a large, shallow, eutrophic lake, Lake Taihu, China[J]. Environ. Pollut. 157(8-9), 2507-2513.

[3] Brandes J A, Ingall E, Paterson D. 2007. Characterization of minerals and organic phosphorus species in marine sediments using soft X-ray fluorescence spectromicroscopy[J]. Mar. Chem. 103(3-4), 250-265.

[4] Brooker M R, Longnecker K, Kujawinski E B, et al. 2018. Discrete Organic Phosphorus Signatures are Evident in Pollutant Sources within a Lake Erie Tributary[J]. Environ. Sci. Technol. 52(12), 6771-6779.

[5] Buenemann E K, Smernik R J, Doolette A L, et al. 2008. Forms of phosphorus in bacteria and fungi isolated from two Australian soils[J]. Soil. Biol. Biochem. 40(7), 1908-1915.

[6] Cao X, Wang Y, He J, et al. 2016. Phosphorus mobility among sediments, water and cyanobacteria enhanced by cyanobacteria blooms in eutrophic Lake Dianchi[J]. Environ. Pollut. 219, 580-587.

[7] Chang D, Zhang Y. 2023. Farmland nutrient pollution and its evolutionary relationship with plantation economic development in China[J]. J. Environ. Manage. 325(Pt B), 116589.

[8] Chen M S, Ding S M, Wu Y X, et al. 2019. Phosphorus mobilization in lake sediments: Experimental evidence of strong control by iron and negligible influences of manganese redox reactions[J]. Environ. Pollut. 246, 472-481.

[9] Chen W, He C, Gu Z, et al. 2020a. Molecular-level insights into the transformation mechanism for refractory organics in landfill leachate when using a combined semi-aerobic aged refuse biofilter and chemical oxidation process[J]. Sci. Total Environ. 741, 140502.

[10] Chen W, He C, Zhuo X, et al. 2020b. Comprehensive evaluation of dissolved organic matter molecular transformation in municipal solid waste incineration leachate. Chem. Eng. J. 400.

[11] Chen W, Wang F, He C, et al. 2020c. Molecular-level comparison study on microwave irradiation-activated persulfate and hydrogen peroxide processes for the treatment of refractory organics in mature landfill leachate[J]. J. Hazard.

Mater. 397, 122785.

[12] Chiwa M. 2020. Ten-year determination of atmospheric phosphorus deposition at three forested sites in Japan[J]. Atmospheric Environ. 223, 117247.

[13] Choi H J, Choi C H, Lee S M. 2009. Analyses of phosphorus in sewage by fraction method[J]. J. Hazard. Mater. 167(1-3), 345-350.

[14] Darch T, Blackwell M S A, Hawkins J M B, et al. 2014. A Meta-Analysis of Organic and Inorganic Phosphorus in Organic Fertilizers, Soils, and Water: Implications for Water Quality[J]. Crit. Rev. Environ. Sci. Technol. 44(19), 2172-2202.

[15] Du F, Hua L, Zhai L, et al. 2022a. Rice-crayfish pattern in irrigation-drainage unit increased N runoff losses and facilitated N enrichment in ditches[J]. Sci. Total Environ. 848, 157721.

[16] Du Y, An S, He H, Wen S, et al. 2022b. Production and transformation of organic matter driven by algal blooms in a shallow lake: Role of sediments[J]. Water Res. 219, 118560.

[17] Duhamel S, Diaz J M, Adams J C, et al. 2021. Phosphorus as an integral component of global marine biogeochemistry[J]. Nat. Geosci. 14(6), 359-368.

[18] Feng W, Yang F, Zhang C, et al. 2020. Composition characterization and biotransformation of dissolved, particulate and algae organic phosphorus in eutrophic lakes[J]. Environ. Pollut. 265(Pt B), 114838.

[19] Gao S X, Zhang X, Fan W Y, et al. 2021. Molecular insight into the variation of dissolved organic phosphorus in a wastewater treatment plant[J]. Water Res. 203, 117529.

[20] Hafuka A, Tsubokawa Y, Shinohara R, et al. 2021. Phosphorus compounds in the dissolved and particulate phases in urban rivers and a downstream eutrophic lake as analyzed using ^{31}P NMR[J]. Environ. Pollut. 288, 117732.

[21] Huang J, Xu C, Ridoutt B G, et al. 2017. Nitrogen and phosphorus losses and eutrophication potential associated with fertilizer application to cropland in China[J]. J. Clean. Prod. 159, 171-179.

[22] Huang J, Zhang Y, Bing H, et al. 2021. Characterizing the river water quality in China: Recent progress and on-going challenges[J]. Water Res. 201, 117309.

[23] Huang L, Han X, Wang X, et al. 2022. Coupling with high-resolution remote

sensing data to evaluate urban non-point source pollution in Tongzhou, China[J]. Sci. Total Environ. 831, 154632.

[24] Huser B J, Pilgrim K M. 2014. A simple model for predicting aluminum bound phosphorus formation and internal loading reduction in lakes after aluminum addition to lake sediment[J]. Water Res. 53, 378-385.

[25] Jia Y, Sun S, Wang S, et al. 2022. Phosphorus in water: A review on the speciation analysis and species specific removal strategies[J]. Crit. Rev. Environ. Sci. Technol. 53(4), 435-456.

[26] Jónasdóttir S H. 2019. Fatty Acid Profiles and Production in Marine Phytoplankton[J]. Mar. Drugs 17(3), 151.

[27] Kellerman A M, Kothawala D N, Dittmar T, et al. 2015. Persistence of dissolved organic matter in lakes related to its molecular characteristics[J]. Nat. Geosci. 8(6), 454-457.

[28] Kruse J, Leinweber P, Eckhardt K U, et al. 2009. Phosphorus $L_{2,3}$-edge XANES: overview of reference compounds[J]. J. Synchrotron Radiat. 16(Pt 2), 247-259.

[29] Kurek M R, Harir M, Shukle J T, et al. 2021a. Seasonal transformations of dissolved organic matter and organic phosphorus in a polymictic basin: Implications for redox-driven eutrophication[J]. Chem. Geol. 573, 120212.

[30] Kurek M R, Harir M, Shukle J T, et al. 2021b. Seasonal transformations of dissolved organic matter and organic phosphorus in a polymictic basin: Implications for redox-driven eutrophication[J]. Chem. Geol. 573.

[31] Li J C, Chen Y, Cai K K, et al. 2022a. A high-resolution nutrient emission inventory for hotspot identification in the Yangtze River Basin[J]. J. Environ. Manage. 321, 115847.

[32] Li W, Li B, Tao S, et al. 2020. Missed atmospheric organic phosphorus emitted by terrestrial plants, part 2: Experiment of volatile phosphorus[J]. Environ. Pollut. 258, 113728.

[33] Li W, Li B, Tao S, et al. 2021. Source identification of particulate phosphorus in the atmosphere in Beijing. Sci. Total Environ. 762, 143174.

[34] Li X, Guo M, Wang Y, et al. 2022b. Molecular insight into the release of phosphate from dissolved organic phosphorus photo-mineralization in shallow lakes based on FT-ICR MS analysis[J]. Water Res. 222, 118859.

[35] Li Y, Wang H, Deng Y, et al. 2022c. How climate change and land-use evolution relates to the non-point source pollution in a typical watershed of China[J]. Sci. Total Environ. 839, 156375.

[36] Liu Q, Liu S, Zhao H, et al. 2015. The phosphorus speciations in the sediments up- and down-stream of cascade dams along the middle Lancang River[J]. Chemosphere 120, 653-659.

[37] Lusk M G, Toor G S. 2016. Dissolved organic nitrogen in urban streams: Biodegradability and molecular composition studies[J]. Water Res. 96, 225-235.

[38] Ma X, Jiao X, Sha Z, et al. 2022. Characterization of atmospheric bulk phosphorus deposition in China[J]. Atmospheric Environ. 279, 119127.

[39] Newman E I. 1995. Phosphorus inputs to terrestrial ecosystems[J]. J. Ecol. 83(4), 713-726.

[40] Ni Z, Li Y, Wang S. 2022. Cognizing and characterizing the organic phosphorus in lake sediments: Advances and challenges[J]. Water Res. 220, 118663.

[41] Ni Z, Wang S, Wu Y, et al. 2020. Response of phosphorus fractionation in lake sediments to anthropogenic activities in China[J]. Sci. Total Environ. 699, 134242.

[42] Ni Z, Wang S, Zhang, B T, et al. 2019. Response of sediment organic phosphorus composition to lake trophic status in China[J]. Sci. Total Environ. 652, 495-504.

[43] Niraula R, Kalin L, Srivastava P, et al. 2013. Identifying critical source areas of nonpoint source pollution with SWAT and GWLF[J]. Ecol. Modell. 268, 123-133.

[44] Ohno T, Parr T B, Gruselle M C I, et al. 2014. Molecular composition and biodegradability of soil organic matter: A case study comparing two new England forest types[J]. Environ. Sci. Technol. 48(13), 7229-7236.

[45] Okin G S, Mahowald N, Chadwick O A, et al. 2004. Impact of desert dust on the biogeochemistry of phosphorus in terrestrial ecosystems[J]. Global Biogeochem. Cycles 18(2), GB2005.

[46] Oldfield L, Rakhimbekova S, Roy J W, et al. 2020. Estimation of phosphorus

loads from septic systems to tributaries in the Canadian Lake Erie Basin[J]. Journal of Great Lakes Research 46(6), 1559-1569.

[47] Ongley E D, Xiaolan Z, Tao Y. 2010. Current status of agricultural and rural non-point source Pollution assessment in China[J]. Environ. Pollut. 158(5), 1159-1168.

[48] Peng Y, Tian C, Chi M, Yang H. 2019. Distribution of phosphorus species and their release risks in the surface sediments from different reaches along Yellow River[J]. Environ. Sci. Pollut. Res. Int. 26(27), 28202-28209.

[49] Pu J, Ni Z, Wang S. 2020. Characteristics of bioavailable phosphorus in sediment and potential environmental risks in Poyang Lake: The largest freshwater lake in China[J]. Ecol. Indic. 115, 106409.

[50] Pu J, Wang S, Fan F, et al. 2023. Recognizing the variation of bioavailable organic phosphorus in sediment and its significance between high and low risk periods for algal blooms in Lake Erhai[J]. Water Res. 229, 119514.

[51] Saha A, Vijaykumar M E, Das B K, et al. 2023. Geochemical distribution and forms of phosphorus in the surface sediment of Netravathi-Gurupur estuary, southwestern coast of India[J]. Mar. Pollut. Bull. 187, 114543.

[52] Shao M Q, Zhao G, Kao S C, et al. 2020. Quantifying the effects of urbanization on floods in a changing environment to promote water security - A case study of two adjacent basins in Texas[J]. J. Hydrol. 589, 125154.

[53] Shen Y, Benner R. 2020. Molecular properties are a primary control on the microbial utilization of dissolved organic matter in the ocean[J]. Limnol. Oceanogr. 65(5), 1061-1071.

[54] Shrestha S, Bhatta B, Shrestha M, et al. 2018. Integrated assessment of the climate and landuse change impact on hydrology and water quality in the Songkhram River Basin, Thailand[J]. Sci. Total Environ. 643, 1610-1622.

[55] Tao P, Huang T, Sun T, et al. 2022. Recycling of internal phosphorus during cyanobacterial growth and decline in a eutrophic lake in China indicated by phosphate oxygen isotopes[J]. Appl. Geochemistry 141, 105320.

[56] van der Veen I, de Boer J. 2012. Phosphorus flame retardants: Properties, production, environmental occurrence, toxicity and analysis[J]. Chemosphere 88(10), 1119-1153.

[57] Wang Y H, Yang Y N, Liu X, et al. 2021. Interaction of Microplastics with Antibiotics in Aquatic Environment: Distribution, Adsorption, and Toxicity. Environ[J]. Sci. Technol. 55(23), 15579-15595.

[58] Xie Z, Ye C, Li C, et al. 2022. The global progress on the non-point source pollution research from 2012 to 2021: a bibliometric analysis[J]. Environ. Sci. Eur. 34(121), 1-17.

[59] Xue J, Wang Q, Zhang M. 2022. A review of non-point source water pollution modeling for the urban-rural transitional areas of China: Research status and prospect[J]. Sci. Total Environ. 826, 154146.

[60] Yang B, Zhou J B, Lu D L, et al. 2019. Phosphorus chemical speciation and seasonal variations in surface sediments of the Maowei Sea, northern Beibu Gulf[J]. Mar. Pollut. Bull. 141, 61-69.

[61] Zhang T, Yang Y H, Ni J P, et al. 2019. Adoption behavior of cleaner production techniques to control agricultural non-point source pollution: A case study in the Three Gorges Reservoir Area[J]. J. Clean. Prod. 223, 897-906.

[62] Zhao B, Hu Y, Yu H, et al. 2023. A method for researching the eutrophication and N/P loads of plateau lakes: Lugu Lake as a case[J]. Sci. Total Environ. 876, 162747.

[63] Zhou Y P, Zhao C, He C, et al. 2022. Characterization of dissolved organic matter processing between surface sediment porewater and overlying bottom water in the Yangtze River Estuary[J]. Water Res. 215.

[64] Zhu Z, Wang Z, Yu Y, et al. 2023. Occurrence forms and environmental characteristics of phosphorus in water column and sediment of urban waterbodies replenished by reclaimed water[J]. Sci. Total Environ. 888, 164069.

[65] Zou L, Liu Y, Wang Y, et al. 2020. Assessment and analysis of agricultural non-point source pollution loads in China: 1978-2017[J]. J. Environ. Manage. 263, 110400.

[66] 付超, 苏晶, 赵海萍, 等. 2020. 基于GIS的漳河上游城市非点源污染负荷估算[J]. 水资源保护, 36(3), 60.

第 3 章

流域磷的迁移转化

本章基于前文对磷来源与形态的总结，进一步阐述不同来源不同形态中的磷在流域中的迁移转化过程，包括运输、沉积、转化与释放等，分析这些磷形态对流域的具体影响。由于磷在流域中的迁移转化与其源解析密切相关，总结不同磷形态的环境效应与迁移转化机制，可为其定性源识别与定量源解析提供理论依据与数据支撑。

3.1 磷源的组成特征

前面章节分别总结了磷的主要来源和流域中存在的主要磷形态，本节将二者联系起来，阐述不同磷源的磷形态组成特征，从而进一步引出后文对磷迁移转化的讨论。不同来源中的磷均有特异性组成（图3-1）。从物理形态看，WTTP处理后的磷以 SRP 为主（Millier, Hooda, 2011）。农业径流中的磷以 POP、DOP 和 DIP 为主（Sutherland, Bramucci, 2022; Yuan et al., 2021）；从化学组成看，岩石风化和磷矿开采中的磷以 Ca-P 为主（Ni et al., 2020）。生活污水和工业废水中磷的主要成分是 Fe/Al-P（Huang et al., 2021; Martin et al., 2020; Yuan et al., 2023）。而畜禽养殖和肥料中磷的主要成分是 Hum-P$_o$ 和 Res-P$_o$（Ni et al., 2020）。

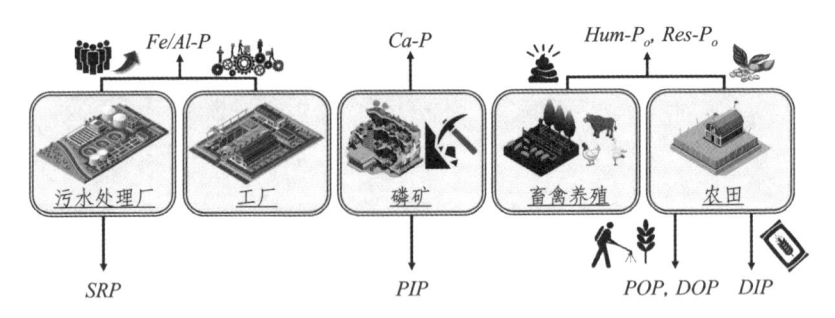

图 3-1　不同污染源的磷形态组成特征

磷的形态组成不同是由不同的物化过程和人类活动造成的。岩石、土壤的风化以及磷矿开采使得 Ca-P 进入流域（Ni et al., 2020）；人口的增长和工业的发展导致工业和生活污水排放持续加剧，工业废水和生活污水中 Fe、Al 和磷酸盐含量较高，这使得废水中大量的 Fe/Al-P 进入流域（Huang et al., 2021; Martin et al., 2020; Yuan et al., 2023）；农业的集约化加剧了化肥和畜禽养殖的磷负荷。农田未利用肥料是农业径流 DIP 的主要来源，而 POP、DOP 则是来源于农药和植物残体（Sutherland, Bramucci, 2022; Yuan et al., 2021）。Hum-P$_o$ 和 Res-P$_o$ 是含磷物质在豆类、谷物和非反刍动物不消化粪便中的主要存在形式,这两种磷形态在畜禽养殖废水中普遍存在(Ni et al., 2020)。

3.2 磷循环

磷形态与磷循环密不可分,流域中的磷循环是一个复杂的过程,包括了磷的运输与沉积、水体中磷的转化以及沉积物中磷的释放。这个过程牵涉多种磷形态的相互转化和迁移。

3.2.1 磷的运输与沉积

通过人类活动、大气沉降、降水和岩石风化等过程输入流域的磷会通过稀释扩散,平流运输等方式在流域中运输,并在水流速度减缓的地方沉积(如河床、湖泊和河口等)(Xue et al., 2022; Zou et al., 2020)。磷通常存在四种沉积方式:① 吸附在黏土上;② 破碎磷酸盐沉积;③ 磷灰石沉积;④ 钙沉积(Song and Song, 2019)。其中,正磷酸盐吸附在沉积物上的机制包括离子交换、配体交换、氢键和表面沉淀等(图3-2)(Jia et al., 2022)。此外,P_o也可以通过配体交换吸附在矿物质上(Ni et al., 2022)。流域中的磷会在水体、悬浮颗粒物和沉积物中广泛存在,由此便牵扯到水体中磷的相互转化以及沉积物中磷的释放。

3.2.2 水体中磷转化

水体中的磷是以PP和DP两种形式存在的,在复杂的物理化学条件下两者之间会发生积极转化。悬浮颗粒物(Suspended Partice,SP)可充当临时中转站,是固-液界面磷转化的汇(Pu et al., 2021)。总颗粒磷(Total Particulate Phosphorus,TPP)和总溶解磷(Total Dissolved Phosphorus,TDP)比例变化会影响富营养化进程(Pu et al., 2021)。根据Pu et al.(2021)的研究SP作为水体中磷的汇,可以调控TDP和TPP的比例。

具体地,如图3-3所示,通过分配系数(partition coefficient,K_P)来评价颗粒的吸附能力(Galunin et al., 2014)。该参数曾被用于量化水环境中的分配吸收量(Lin et al., 2012; Lin et al., 2013)。SRP和TDP的吸附量与平衡吸附浓度之间的关系拟合为抛物线曲线如图3-3(a)和(b)所示,呈现出"先增后降"的趋势,在400 mg/L的SP水平下,SRP和TDP的吸附量最大,分别为64.54和62.23 mg/kg。这表明低SP水平有利于磷的吸附和絮凝,而高SP水平(>400 mg/L)会降低磷的吸附能力。在较低的 SP 水平下，SRP 和 TDP 的 K_P 急剧增加，随后在较高的 SP 水平下趋于稳定。在较高的 SP 水平下（>600 mg/L），SRP 的 K_P 继续增加，而 TDP 的 K_P 降低。

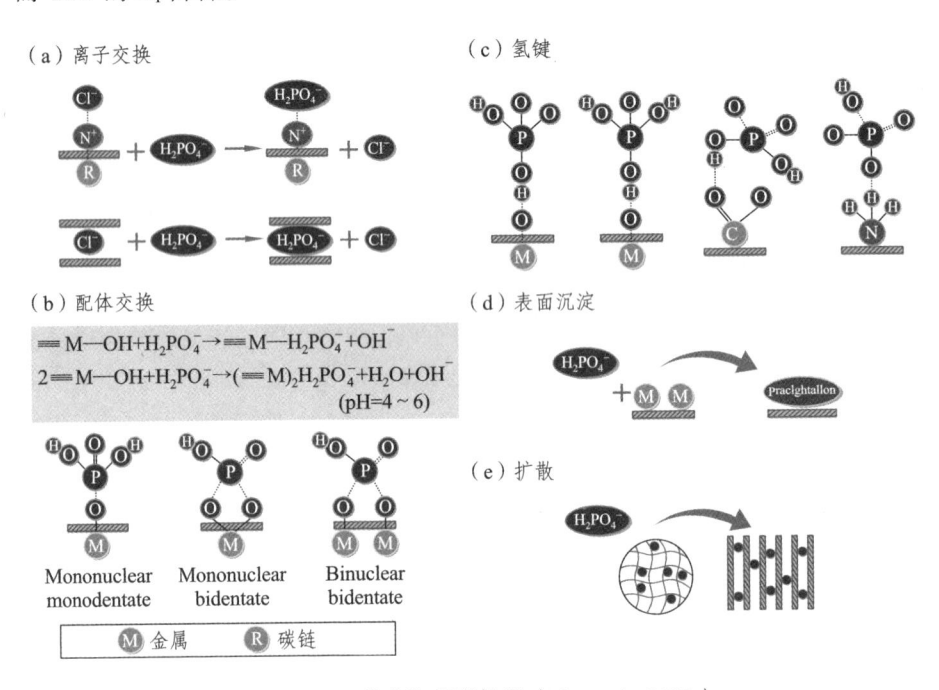

图 3-2　正磷酸盐吸附机理（Jia et al., 2022）

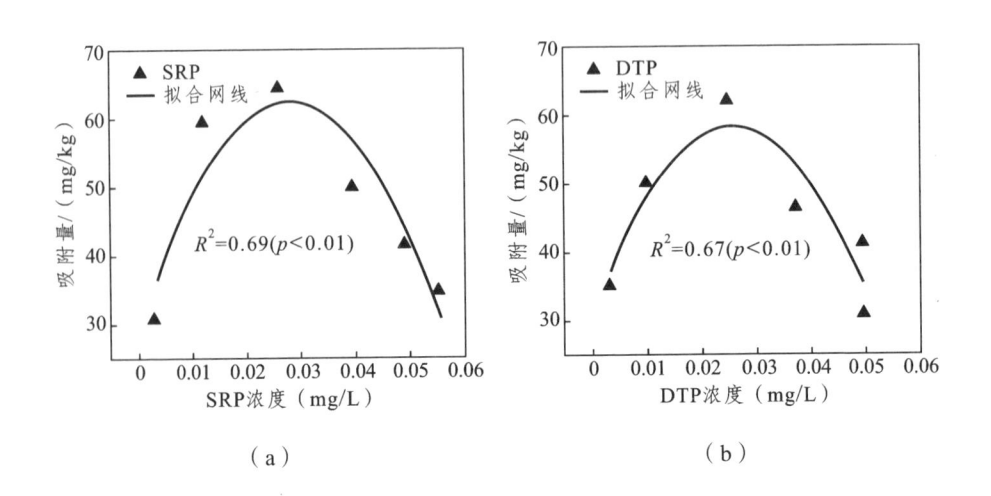

（a）
（b）

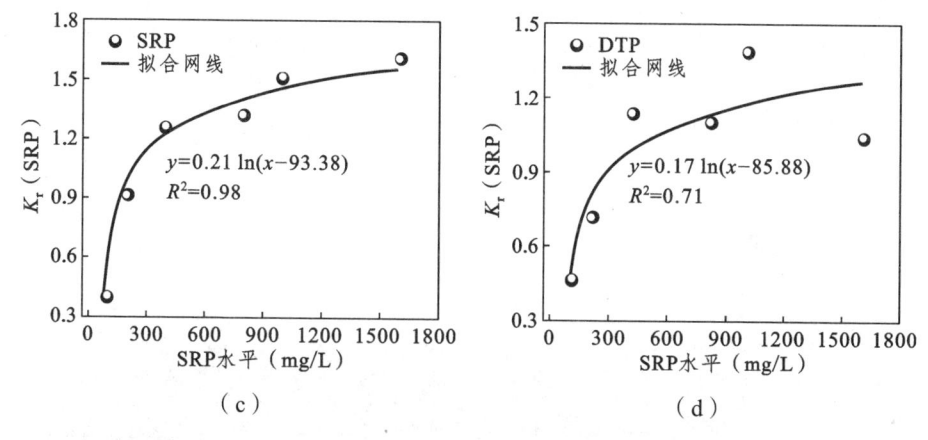

(c) 图中: $y=0.21\ln(x-93.38)$, $R^2=0.98$, 纵轴 K_r(SRP), 横轴 SRP 水平 (mg/L)

(d) 图中: $y=0.17\ln(x-85.88)$, $R^2=0.71$, 纵轴 K_r(SRP), 横轴 SRP 水平 (mg/L)

图 3-3 SRP（a）和 TDP（b）的吸附模型；SRP（c）和 TDP（d）
在不同 SP 水平上的分配系数（Pu et al., 2021）

同样地，P_i 和 P_o 之间也在生物化学作用下发生积极的相互转化。P_i 可通过生物的同化作用转化为 P_o，而 P_o 可以通过不同的机制转化为生物可利用磷（Bai et al., 2017）。当 ortho-P 浓度不足以满足初级生产需要时，藻类和细菌可以通过 APA 有效利用 P_o 来维持生长（Ni et al., 2022）；除酶水解外，活性氧物种（Reactive Oxygen Species，ROS）是 P_o 光矿化的另一可行途径。ROS 是指来自氧的自由基和非自由基，包含了超氧阴离子（O_2^-）、过氧化氢（H_2O_2）、羟自由基（·OH）、臭氧（O_3）和单线态氧（$^1O_2$）等，由于它们含有不成对的电子，因而具有很高的化学反应活性，可以矿化难降解有机物（Chen et al., 2020; Hu, 2021; Hu et al., 2022）。溶解性有机质（dissolved organic matter）很容易在光催化下产生 ROS（Chen and Li, 2020）。在 ROS 的作用下，高分子量和高双键的 P_o 优先被去除，·OH 是 P_o 光矿化的主要驱动力（Li et al., 2022）。以上两种途径将释放 P_i 用于水生植物和浮游动物的生长。而当浮游动物死亡时，又将释放 P_o 回到水体中。Feng et al.（2018）研究表明，浮游植物碎屑水解的 P_o 的比例大约是水生植物水解 P_o 的 4 倍，是沉积物水解 P_o 的 25 倍，这部分 P_o 以 Mono-P 为主。

3.2.3 沉积物磷释放

沉积物既流域中磷的汇，也是重要的磷源，参与磷的释放。沉积物存在的内源磷对流域的贡献不容忽视。Zhao et al.（2023）使用内源静态释放实验和外源改进输出系数模型分别计算了流域内源和外源氮磷营养负荷，其中来自沉积物释放的磷负荷占总负荷的比例达到了 57.4%。如图 3-4 所示，水力扰动下的再悬浮是沉积物中磷进入水体最直接的方式。除此之外，对于 P_i，其释放是由固-液界面的吸附/解吸，沉淀/溶解反应控制，浮游动物和水生植物介导，并受理化因素如 pH、温度、DO 和氧化还原条件等影响（Dieter et al., 2015; Pu et al., 2021）。

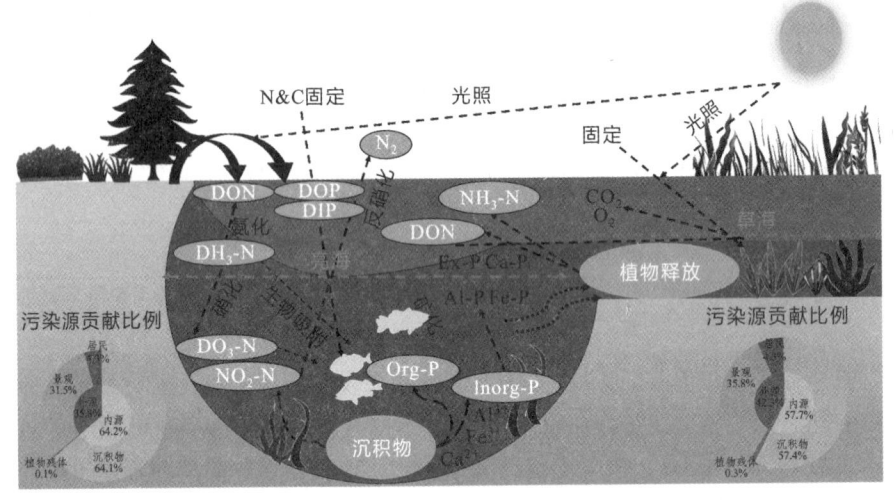

图 3-4 流域磷负荷占比与流域磷释放机理

Fe-P 是沉积物中最活跃的无机磷形态，被认为是衡量沉积物中磷释放潜能的重要指标。DO 和 pH 是决定 Fe-P 沉淀和溶解平衡的关键因素，低氧还原环境下将促进其溶解（Chen et al., 2019; Martins et al., 2014; Pu et al., 2021; Wang et al., 2023; Wang et al., 2019）。同时，藻类等水生植物将介导沉积物中磷的释放。具体地，一方面，藻华期间沉积物-水界面缺氧条件的形成，引起 Fe(Ⅲ)氢氧化物还原性溶解，导致 Fe-P 释放；另一方面，藻类光合使水体和上层沉积物 pH 升高，促进 OH^- 与 PO_4^{3-} 竞争吸附，导致 PO_4^{3-} 从沉积物中解吸。藻华驱动沉积物中 P 释放，P 释放反过来促进水华生长，形成正反馈（Ni et al., 2022）。Ca-P 通常被认为是沉积物中相对稳定的组分。仅在水生植物发达的沉积物中，Ca-P 由水生植物介导，从而释放到水体中。植物根系渗出的有机酸可以降低根际 pH 值，促进 Ca-P 的释放（Li et al., 2021; Liu et al., 2022; Zhu et al., 2023）。

对于 P_o，可以通过生物矿化和化学分解过程释放 DP（Ni et al., 2022）。Mono-P、Diester-P 和 DNA 等 P_o 可以通过 APA 水解为生物可利用的 ortho-P（Feng et al., 2018; Ni et al., 2020）。在生物酶的作用下，ortho-P 可以很频繁地与沉积物中的 P_o 发生交换，进一步促进 P_o 矿化（Hafuka et al., 2021; Li et al., 2019; Zhu et al., 2023）；光解是沉积物中 P_o 释放的另一重要途径，水中天然存在的光敏剂 NO_3^-、

Fe^{3+} 和溶解性有机质等产生的 ROS 在光照下加速了 P_o 的降解（Jiang et al., 2016）。Li et al.（2019）研究表明，由于表层沉积物中有机物和·OH 含量高，P_i 的释放量也随之增加。

基于上述关于磷来源、磷形态及其迁移转化的讨论，最终在图 3-5 中构建了流域全过程磷循环。值得注意的是，与氮循环不同，磷在流域中很难形成闭合的循环，最终大部分的磷都会流失，一定程度上造成了资源的浪费（Wang et al., 2022）。因此明晰磷在流域中的转化规律与途径，对追踪磷来源以及保持磷资源均有十分重要的意义。

如图 3-5 所示，不同污染源（点源与面源）具有不同的磷形态组成特征。这些不同形态的磷进入流域后，通过扩散、平流运输等方式在流域中迁移，最终在水流速度减缓的地方沉积（如河床、湖泊和河口等）。在沉积物中，Ex-P 由于结合不紧密，在水动力扰动下很容易发生再悬浮重新回到水体中。Fe-P 极易在高 pH 的还原环境下释放，Ca-P 则通过水生植物分泌的有机酸溶解释放；在水体中，DP 和 PP 在 SP 的介导下会发生积极的相互转化。P_i 和 P_o 也在藻类等水生动植物的介导下相互转化。其中 P_o 可通过酶解和光解两种途径转化为 P_i。上述迁移转化过程所释放的 Pi（以 Ortho-P 为主）易被藻类吸收，从而引起藻华。

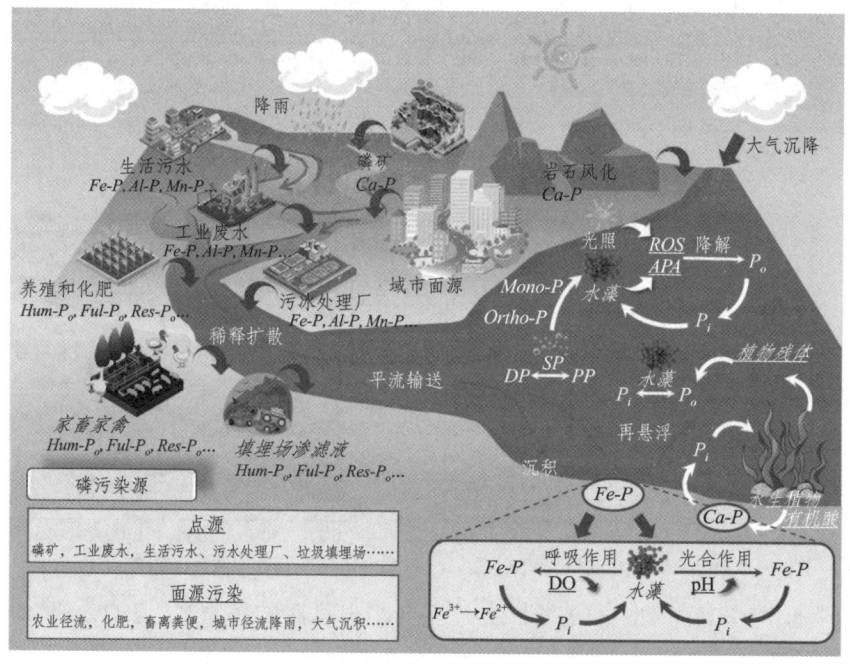

图 3-5 流域磷污染来源与流域全过程磷循环

3.3 磷形态对流域的影响

根据前文的论述，在流域中磷有多种赋存形式。经过复杂的迁移转化过程后，不同的赋存形式可能会对流域产生不同的影响。为了具体分析不同磷形态对流域富营养化的影响，我们根据近 5 年（2019—2023 年）关于流域磷形态的研究，总结了磷形态对流域水体 TP 和叶绿素 a（Chl-a）的影响，如图 3-6 所示（彩图见附录 2）。

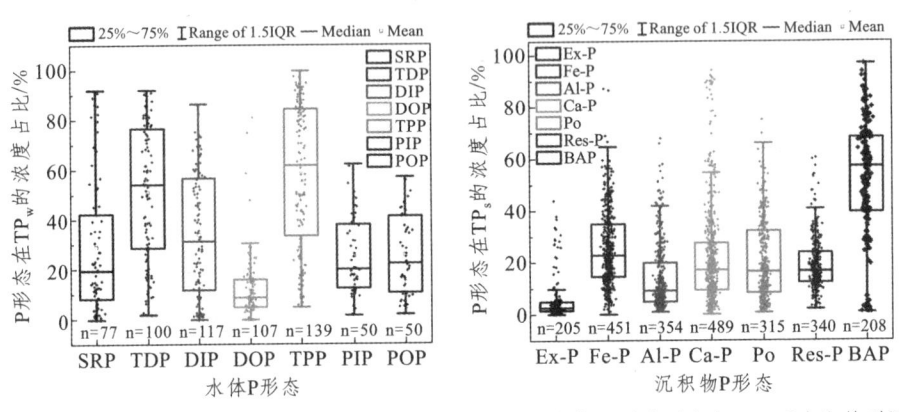

（a）水体中不同磷形态占 TP_w 百分比箱型图　（b）沉积物中不同磷形态占 TP_s 百分比箱型图

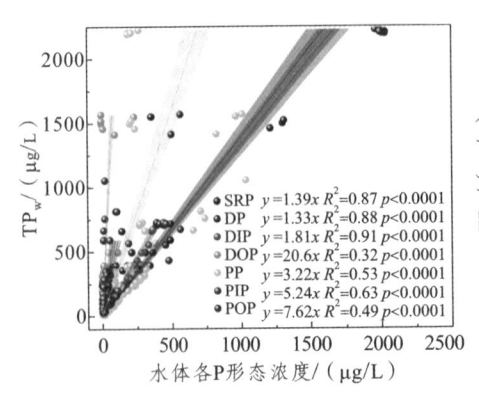

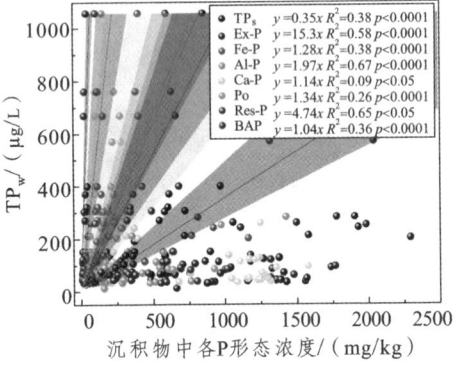

（c）水体中不同磷形态与 TP_w 线性回归　（d）沉积物中不同磷形态与 TP_w 线性回归

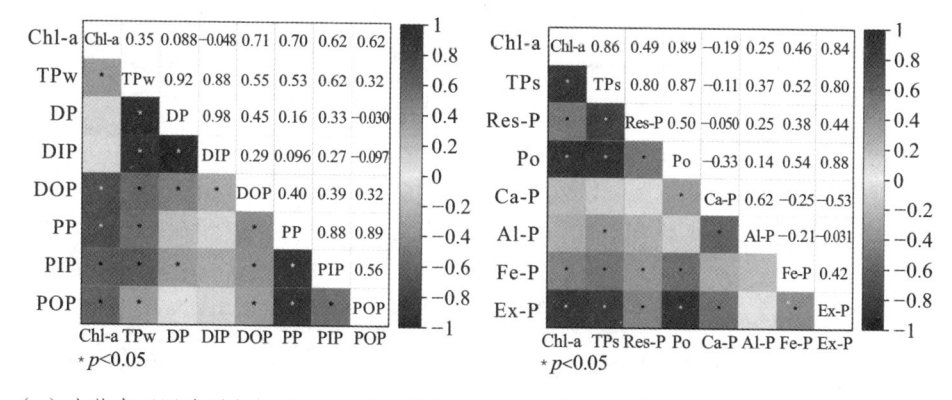

(e) 水体中不同磷形态与 Chl-a 的相关性分析 (f) 沉积物中不同磷形态与 Chl-a 的相关性分析

图 3-6 不同磷形态对水体 TP 与 Chl-a 的影响分析

如图 3-6（a）和（b）所示，无论是在水体（如 TP_w）还是沉积物（如 TP_s）中，不同形态的磷对 TP 占比均不同，且占比跨度很大。如 DIP、Fe-P 和 P_o 这一类较为活跃磷形态占比相对高。如图 3-6（c）和（d）所示，不同的磷形态与 TP_w 的 R^2 值有较大差别。在水体中，SRP 和 DIP 与 TP_w 的 R^2 分别为 0.87 和 0.91，表明这两种形态的磷与 TP_w 相关性较强。在沉积物中，TP_s 与 TP_w 的 R^2 值为 0.38，而 Ex-P（$R^2=0.58$, $p<0.0001$）和 Al-P（$R^2=0.67$, $p<0.0001$）与 TP_w 的相关性更强。Ex-P 和 Al-P 均为沉积物中较为活跃的磷形态，容易向水体中释放（Pu et al., 2020）。以上结果表明，水体中不同磷形态对总磷的贡献有较大差异，同时沉积物对水体中的磷赋存情况影响也较大。

Chl-a 浓度通常用于表征水体富营养化程度（Du et al., 2022），为评价磷形态对流域富营养的影响，分别构建 Chl-a 与水体和沉积物中磷形态的 pearson 相关性分析，如图 3-6（e）和（f）所示。在水体中，与 Chl-a 相关性最强的不是 TP_w，而是 DOP、POP 和 PIP 等形态。DOP 可以通过碱性磷酸酶（Alkaline Phosphatase, APA）降解为生物利用度更高的正磷酸盐（ortho-P, PO_4^{3-}）（Feng et al., 2018; Ni et al., 2020）。光解是 DOP 降解的另一重要途径，光照下，水中天然存在的光敏剂 NO_3^-、Fe^{3+} 和 DOM 等产生的 ROS 同样会加速 DOP 的降解（Li et al., 2019）。Pu et al.（2021）表明 SP 可充当临时中转站，是固-液界面 P 转化的汇。SP 作为水体中磷的汇，可以调控 DP 和 PP 的比例，从而影响磷形态的生物利用性（Pu et al., 2021）。由此，DOP 和 PP 是对水体富营养化具有显著贡献的磷形态。

在沉积物中，TP_s 与 Chl-a 显著相关（$p<0.05$）。其中，Ex-P，Fe-P 和 P_o 与 Chl-a 相关性较强，这三者均属于 BAP。根据前文的描述，Ex-P 磷通常松散吸附在沉积物表面，很容易进入水体被初级生产者直接吸收（Pu et al., 2020; Saha et al., 2023）；Fe-P 是沉积物中最活跃的组分，在还原环境下会大量释放到上覆水中（Choi et al., 2009; Ni et al., 2022）；沉积物中 P_o 同样可由 APA 和 ROS 两种途径矿化，从而提高生物利用性（Feng et al., 2018）。Ca-P 与 Chl-a 相关性较弱，在图 3-6（d）中，Ca-P 与 TP_w 相关性同样较弱（$R^2=0.09, p<0.05$）。Ca-P 以磷灰石为主，通常被认为是沉积物中相对稳定的组分，仅在水生植物发达的沉积物中，由植物根系渗出的有机酸，促进其释放（Li et al., 2021; Liu et al., 2022; Zhu et al., 2023）。上述磷形态的分析表明，磷对流域富营养化的影响高度依赖于不同形态磷的化学组成，而不是 TP_w 的绝对浓度。

参考文献

[1] Bai X L, Sun, J H, Zhou, Y K, et al. 2017. Variations of different dissolved and particulate phosphorus classes during an algae bloom in a eutrophic lake by ^{31}P NMR spectroscopy[J]. Chemosphere 169, 577-585.

[2] Chen M S, Ding S M, Wu Y X, et al. 2019. Phosphorus mobilization in lake sediments: Experimental evidence of strong control by iron and negligible influences of manganese redox reactions[J]. Environ. Pollut. 246, 472-481.

[3] Chen W, Li Q. 2020. Elimination of UV-quenching substances from MBR- and SAARB-treated mature landfill leachates in an ozonation process: A comparative study[J]. Chemosphere 242, 125256.

[4] Chen W, Wang F, He C, et al. 2020. Molecular-level comparison study on microwave irradiation-activated persulfate and hydrogen peroxide processes for the treatment of refractory organics in mature landfill leachate[J]. J. Hazard. Mater. 397, 122785.

[5] Choi H J, Choi C H, Lee S M. 2009. Analyses of phosphorus in sewage by fraction method[J]. J. Hazard. Mater. 167(1-3), 345-350.

[6] Dieter D, Herzog C, Hupfer M. 2015. Effects of drying on phosphorus uptake in re-flooded lake sediments[J]. Environ. Sci. Pollut. Res. Int. 22(21), 17065-17081.

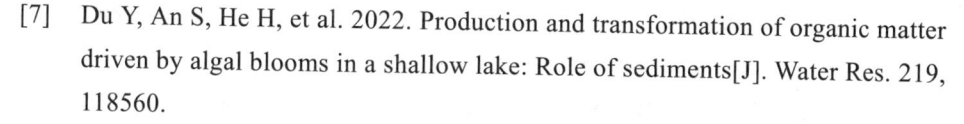

[7] Du Y, An S, He H, et al. 2022. Production and transformation of organic matter driven by algal blooms in a shallow lake: Role of sediments[J]. Water Res. 219, 118560.

[8] Feng W, Wu F, He Z, et al. 2018. Simulated bioavailability of phosphorus from aquatic macrophytes and phytoplankton by aqueous suspension and incubation with alkaline phosphatase[J]. Sci. Total Environ. 616-617, 1431-1439.

[9] Galunin E, Ferreti J, Zapelini I, et al. 2014. Cadmium mobility in sediments and soils from a coal mining area on Tibagi River watershed: Environmental risk assessment[J]. J. Hazard. Mater. 265, 280-287.

[10] Hafuka A, Tsubokawa Y, Shinohara R, et al. 2021. Phosphorus compounds in the dissolved and particulate phases in urban rivers and a downstream eutrophic lake as analyzed using ^{31}P NMR[J]. Environ. Pollut. 288, 117732.

[11] Hu Y. 2021. A microwave radiation-enhanced Fe-C/persulfate system for the treatment of refractory organic matter from biologically treated landfill leachate[J]. RSC Adv. 11(47), 29620-29631.

[12] Hu Y, Gu Z, He J, et al. 2022. Novel strategy for controlling colloidal instability during the flocculation pretreatment of landfill leachate[J]. Chemosphere 287(Pt 1), 132051.

[13] Huang S, Xu H, Shang D, et al. 2021. Phosphorus fractions and release factors in surface sediments of a Tailwater River in Xinmi City, a case study[J]. Sustainability 13(10), 5417.

[14] Jia Y, Sun S, Wang S, et al. 2022. Phosphorus in water: A review on the speciation analysis and species specific removal strategies[J]. Crit. Rev. Environ. Sci. Technol. 53(4), 435-456.

[15] Jiang Y C, Kang N X, Zhou Y Y, et al. 2016. The role of Fe(III) on phosphate released during the photo-decomposition of organic phosphorus in deionized and natural waters[J]. Chemosphere 164, 208-214.

[16] Li H, Song C, Yang L, et al. 2021. Phosphorus supply pathways and mechanisms in shallow lakes with different regime[J]. Water Res. 193, 116886.

[17] Li X, Guo M, Duan X, et al. 2019. Distribution of organic phosphorus species in sediment profiles of shallow lakes and its effect on photo-release of phosphate during sediment resuspension[J]. Environ. Int. 130, 104916.

[18] Li X, Guo M, Wang Y, et al. 2022. Molecular insight into the release of phosphate from dissolved organic phosphorus photo-mineralization in shallow lakes based on FT-ICR MS analysis[J]. Water Res. 222, 118859.

[19] Lin P, Chen M, Guo L D. 2012. Speciation and transformation of phosphorus and its mixing behavior in the Bay of St. Louis estuary in the northern Gulf of Mexico. Geochim[J]. Cosmochim. Acta 87, 283-298.

[20] Lin P, Guo L D, Chen M, et al. 2013. Distribution, partitioning and mixing behavior of phosphorus species in the Jiulong River estuary[J]. Mar. Chem. 157, 93-105.

[21] Liu C, Du Y, Zhong J, et al. 2022. From macrophyte to algae: Differentiated dominant processes for internal phosphorus release induced by suspended particulate matter deposition[J]. Water Res. 224, 119067.

[22] Martin N, Ya V, Leewiboonsilp N, et al. 2020. Electrochemical crystallization for phosphate recovery from an electronic industry wastewater effluent using sacrificial iron anodes[J]. J. Clean. Prod. 276, 124234.

[23] Martins G, Peixoto L, Teodorescu S, et al. 2014. Impact of an external electron acceptor on phosphorus mobility between water and sediments[J]. Bioresour. Technol. 151, 419-423.

[24] Millier H K, Hooda P S. 2011. Phosphorus species and fractionation---why sewage derived phosphorus is a problem[J]. J. Environ. Manage. 92(4), 1210-1214.

[25] Ni Z, Li Y, Wang S. 2022. Cognizing and characterizing the organic phosphorus in lake sediments: Advances and challenges[J]. Water Res. 220, 118663.

[26] Ni Z, Wang S, Wu Y, et al. 2020. Response of phosphorus fractionation in lake sediments to anthropogenic activities in China[J]. Sci. Total Environ. 699, 134242.

[27] Pu J, Ni Z, Wang S. 2020. Characteristics of bioavailable phosphorus in sediment and potential environmental risks in Poyang Lake: The largest freshwater lake in China[J]. Ecol. Indic. 115, 106409.

[28] Pu J, Wang S, Ni Z, et al. 2021. Implications of phosphorus partitioning at the suspended particle-water interface for lake eutrophication in China's largest

freshwater lake, Poyang Lake[J]. Chemosphere 263, 128334.

[29] Saha A, Vijaykumar M E, Das B K, et al. 2023. Geochemical distribution and forms of phosphorus in the surface sediment of Netravathi-Gurupur estuary, southwestern coast of India[J]. Mar. Pollut. Bull. 187, 114543.

[30] Song Y, Song S. 2019. Migration and transformation of different phosphorus forms in rainfall runoff in bioretention system[J]. Environ. Sci. Pollut. Res. Int. 26(30), 30633-30640.

[31] Sutherland D L, Bramucci A. 2022. Dissolved organic phosphorus bioremediation from food-waste centrate using microalgae[J]. J. Environ. Manage. 313, 115018.

[32] Wang C, Thielemann L, Dippold M A, et al. 2023. Reductive dissolution of iron phosphate modifies rice root morphology in phosphorus-deficient paddy soils[J]. Soil Biol. Biochem. 177, 108904.

[33] Wang Z, Guo Q, Tian L. 2022. Tracing phosphorus cycle in global watershed using phosphate oxygen isotopes[J]. Sci. Total Environ. 829, 154611.

[34] Wang Z, Huang S, Li D. 2019. Decomposition of cyanobacterial bloom contributes to the formation and distribution of iron-bound phosphorus (Fe-P): Insight for cycling mechanism of internal phosphorus loading[J]. Sci. Total Environ. 652, 696-708.

[35] Xue J, Wang Q, Zhang M. 2022. A review of non-point source water pollution modeling for the urban-rural transitional areas of China: Research status and prospect[J]. Sci. Total Environ. 826, 154146.

[36] Yuan H, Chen P, Liu E, et al. 2023. Terrestrial sources regulate the endogenous phosphorus load in Taihu Lake, China after exogenous controls: Evidence from a representative lake watershed[J]. J. Environ. Manage. 340, 118016.

[37] Yuan X, Krom M D, Zhang M, et al. 2021. Human disturbance on phosphorus sources, processes and riverine export in a subtropical watershed[J]. Sci. Total Environ. 769, 144658.

[38] Zhao B, Hu Y, Yu H, et al. 2023. A method for researching the eutrophication and N/P loads of plateau lakes: Lugu Lake as a case[J]. Sci. Total Environ. 876, 162747.

[39] Zhu Z, Wang Z, Yu Y, et al. 2023. Occurrence forms and environmental characteristics of phosphorus in water column and sediment of urban waterbodies replenished by reclaimed water[J]. Sci. Total Environ. 888, 164069.

[40] Zou L, Liu Y, Wang Y, et al. 2020. Assessment and analysis of agricultural non-point source pollution loads in China: 1978-2017[J]. J. Environ. Manage. 263, 110400.

第 4 章

磷的源解析——清单分析法

从本章开始，将逐个介绍磷源解析的主要方法，包括清单分析法（inventory analysis）、扩散模型法（diffusion models）和受体模型法（receptor models）。其中，清单分析法是指基于物质平衡，建立一定范围内磷的输入和输出清单，进一步推断磷来源的方法。清单分析法又可分为区域清单分析、物质流清单分析和行业清单分析，本章分别介绍三种清单分析方法，并根据当前研究现状对清单分析法的特点进行总结。

4.1 清单分析的建立

清单分析法是指基于物质平衡，在一定范围内磷的输入和输出清单，从而推断磷的来源的方法（Zou et al., 2020）。具体地，在流域中通过清单分析对磷污染进行溯源一般包括以下步骤：① 根据流域内的特点和实际情况，确定磷输入和输出类型（如化肥、农业污染、工业污染、生活污水、畜禽养殖等）；② 收集流域内不同来源类型磷的输入和输出数据（如质量、浓度、空间和时间分布等）；③ 建立物质平衡模型，进行清单分析，以评估不同来源和途径的磷输入贡献率。例如，我国营养物质排放清单（the China Emission Inventory of Nutrients, CEIN）框架的构建过程，如图 4-1 所示。

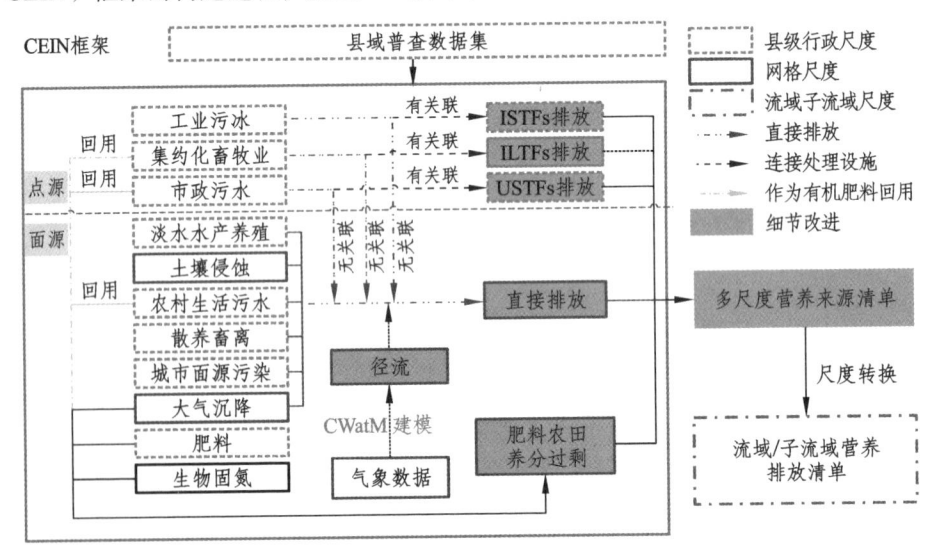

图 4-1 我国营养物质排放清单建立框架

[图中 ISTFs 是工业污水处理设施，USTPs 是城市污水处理厂，ILTFs 是集约化禽畜污水处理设施（Li et al., 2022）]

如图 4-1 所示，CEIN 框架输入空间分辨率为栅格和子流域尺度的营养物质排放数据。基于营养物质的排放数据，进一步量化点源污染负荷，包括城市污水处理厂、工业、集约化畜牧业、淡水水产养殖等，同时也量化面源污染负荷，包括农村生活废水、散养牲畜和农田、城市非点源和土壤侵蚀等（Li et al., 2022）。该框架的优势在于，CEIN 提供 0.1°栅格和子流域尺度的空间分辨率的养分排放数据。我国县的平均面积为 3360 km^2，最小的县面积为 8 km^2。CEIN 的空间分辨率（0.1°，约 11 km^2）能较好地反映县域尺度污染治理措施的效果。县级是我国官方和公共信息披露的最小行政单位。因此，基于县级统计和调查的输入数据（如工业污水、畜禽养殖、生活污水、水产养殖、土壤侵蚀、肥料、大气沉降等），使得精准识别和量化营养物质排放成为可能（Li et al., 2022）。

4.2 区域磷负荷清单分析

区域磷负荷清单分析是指计算一定区域或流域范围内磷的总输入和输出，包括各种来源磷负荷量和分布，以评估该范围内的磷污染程度，并确定污染源和污染途径。例如，Strokal et al.（2016），Bai et al.（2022）和 Sabo et al.（2022）等通过清单分析，量化了长江、珠江和切萨皮克湾流域在一定时间范围（如 1970—2000 年、2016—2020 年和 1985—2019 年）的磷污染输入与输出，从而确定了这些区域的磷污染来源类型。此外，还可以对特定污染物进行清单分析，Pintado-Herrera et al.（2017）表示污水排放是珠江口有机磷阻燃剂（Organophosphorous Flame Retardant, OPFRs）的主要来源。

清单分析法也可用于分析特定海域中营养物质的来源，Yang et al.（2018）根据历史资料，通过清单分析对北黄海西部海域营养物质的长期变化和控制因素进行研究，结果表明北黄海西部流域磷主要来自黄河、鸭绿江和夹江等河流输入以及大气沉降；Zhang et al.（2019）结果表示，黑潮（Kuroshio）是东海营养物质的主要来源，贡献了 84%的 DIP。

近年来，区域磷负荷清单分析逐渐在向大范围、高精度发展，其研究结果通常是在全球或国家范围下，以县域或次流域尺度呈现。Fink et al.（2018）对世界范围内最大的 100 个湖泊磷负荷进行了清单分析，并评估了这些湖泊富营养化情况，结果表明磷引起的湖泊富营养化风险在发展中国家更高，污染来源的主要影响因素是磷肥使用量（Fink et al., 2018）；Sabo et al.（2021）使用次流域尺度上的公开数据（即：磷通量、大气磷沉降、磷需求量和点源排放）编制清单，估算了1945—2001年间累积的 DP 残余，结果表明美国农业的化肥和粪肥利用率不足 50%。

清单分析的数据来源通常是以行政尺度呈现，而模型的建立通常是在水文尺度或非行政尺度上（如流域尺度或网格尺度）。当行政尺度的输入被聚合或分解到水文尺度上时，以行政尺度输入的磷污染重要信息（如人口和农业活动）往往会被删除（Li et al., 2022）。近年来，不少学者提出了多尺度建模方法，该法可以用于克服清单分析法中跨水文和行政尺度建模结果不协调的困难。例如，Chen et al.（2019）开发了一种多尺度的建模方法分析中国河流 DP 来源；Li et al.（2022）在详细调查和多尺度数据转换基础上建立了空间分辨率为 0.1°网格和次流域尺度的营养物质排放清单。

4.3 磷负荷物质流清单分析

磷负荷物质流清单分析评估磷与不同物质之间的相互作用，以及磷在系统中的流动途径和转化过程，以建立物质流清单模型，推断磷的来源。Li et al.（2017）以三阳湿地为研究对象，研究了重金属和营养物质之间的关系；Chen et al.（2008）和 Chen et al.（2010）在排放清单和营养物质全平衡计算的基础上，提出农业磷流模型，描述了中国农业生态系统中的磷流动，如图 4-2 所示。结果表明，2004 年中国农业向地表水排放 TP 达 47.9 万吨，其中畜牧业和矿物肥料的影响最大。

分析热点地区的营养物质流动以及当地农业和社会经济指标，有助于制定特定区域的养分管理技术和政策。Wang et al.（2018）使用营养流动在食物链，环境和资源利用模型（Nutrient flows in Food chains, Environment and Resources use，NUFER）评估了中国粮食生产过程中营养物质的流失，已确定县级尺度上氮磷损失的热点地区。结果表明 1990—2012 年，磷的热点地区扩大了 24 倍，热点覆盖范围不到中国陆地面积的 10%，却造成了超过一半的营养物质流入。其中，动物粪便直接排入河流是造成氮磷流失的重要原因。

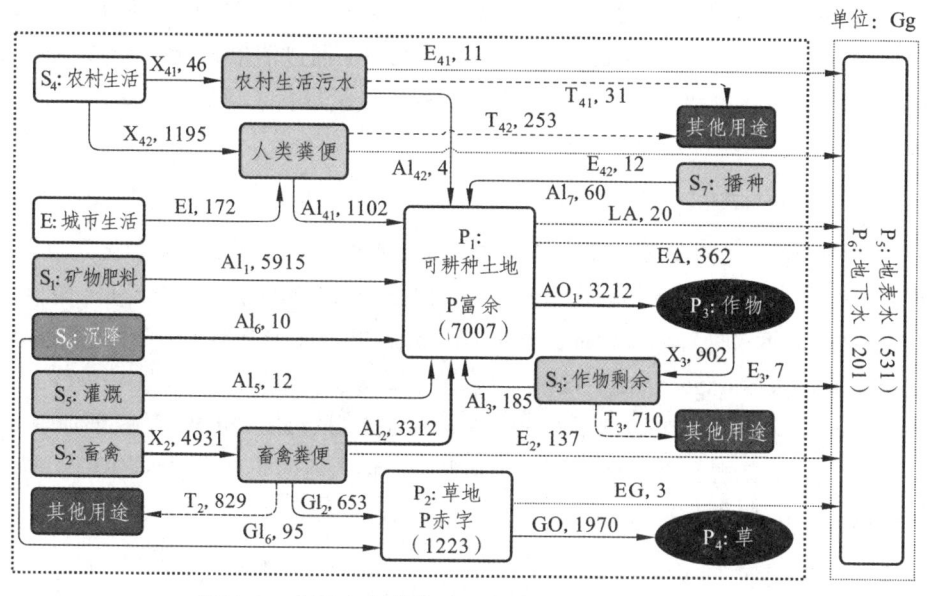

图 4-2　中国农业磷流动示意（Chen et al., 2010）

4.4　行业生产清单分析

1. 农业生产清单分析

农业生产过程中，化肥使用、畜禽饲养和农田灌溉等是流域磷的重要来源，农业生产清单分析是指评估这些农业生产过程中的磷输入和输出，以推断农业磷污染的来源。农业生产清单分析通过对长时间磷的排放清单进行评估，得出农业产业的污染现状。例如：Bouwman et al.（2013）对全球畜禽系统中的磷进行了全面清单分析，发现1950—2000年间，磷过剩翻了44倍；Scherer and Pfister（2015）在5 arc minutes的尺度上，对169种作物进行磷排放清单分析，结果表明当前的研究低估了全球农业磷排放，幅度高达一个数量级。

另外，农业清单分析还可以利用公开的数据，在国家范围内对磷来源的空间与产业分布进行分析。Huang et al.（2017）使用国家数据集分析了中国农业种植系统磷损失，结果表明，中国的海南、云南、广东和福建是全球磷富营养潜力最高的地区。湖南、湖北和四川等产粮大省磷排放对当地的富营养潜力贡献影响显著。Zou et al.（2020）通过清单分析评估了1978—2017年间中国农业面源污染负荷，并在省级尺度上进行了社会经济与时空分析。结果表明，1978—

2017年间TP污染增加了244.1%，畜禽养殖、农村生活垃圾和矿物肥料对总磷贡献达69.1%～88.6%。Chang and Zhang（2023）通过清单分析对1999—2019年间中国9个国家级农业区农田非点源氮磷排放量进行了评价，量化了农业养分污染与人工林经济效应的关系。

2. 城市雨污水排放清单分析

清单分析法可针对城市磷主要来源（如污水处理厂、雨水径流和化粪池等）评估城市雨污水排放对流域磷污染的贡献（Foley et al., 2010）。Smol et al.（2020）对波兰11座污水处理厂（Wastewater Treatment Plants, WWTPs）产生的污水污泥灰（Sewage Sludge Ashes, SSA）进行了清单分析，结果表明这些WWTPs产生的SSA的磷回收潜力达到了1613.8 Mg。Oldfield et al.（2020）估算了加拿大伊利湖流域化粪池系统的磷负荷。结果表明，化粪池向伊利湖输入的磷占总磷负荷的1.7±0.8%～5±2.3%。

3. 工业生产清单分析

磷的工业生产清单分析是指评估特定工业生产过程中磷的输入与输出，包括工艺、设备、技术和废弃物清单等。Luo et al.（2014）和Zhang et al.（2021）研究了工业化养猪场粪便消化方式，量化了环境负担和可持续营养驯化所需的耕地面积，建立了养分排放清单；Herrera et al.（2021）评估了微藻生产设施对环境的影响；He et al.（2021）建立了OPFRs排放和消费清单，表明OPFRs生产过程引起的排放中，占比最高的是纺织品（20.4%）、塑料（16.6%）和橡胶（4.3%）等产业。

4.5 清单分析综合评价

清单分析法综合概述如表4-1所示，清单分析法通过对污染源排放特征、地理分布和社会经济因素的观测和模拟，构建磷污染信息列表，自上而下地核算各污染源排放量。其优势在于，获得的信息范围较大（通常以全球或国家的尺度呈现），时间较长（可达数十年）。近年来，清单分析精度也在不断提升，最小尺度通常为次流域或县域。这有助于决策者制定特定区域的养分管理技术和政策。

表 4-1 磷的源解析-清单分析法综合概述

类型	污染物	最小尺度	调查范围	调查时间段	结果	参考文献
区域清单分析	DIP	次流域	中国	1970—2000	污水是 DIP 的重要来源	(Strokal et al., 2016)
	TP	市域	珠江流域的广东段	2016—2020	主要磷源： 珠江三角洲流域—化肥 珠江三角洲—人类和动物的消费	(Bai et al., 2022)
	TP	国家	切萨皮克湾	1985—2019	磷的主要来源为农业富余、大气沉降和点源负荷	(Sabo et al., 2022)
	OPFRs	流域	珠江三角洲	2012	污水排放是主要污染源，广州和澳门的浓度最高	(Pintado-Herrera et al., 2017)
	PO_4^-	海域	黄海西北部	1976—2006	PO_4^--P 主要来源是黄河、鸭绿江输入和大气沉降	(Yang et al., 2018)
	DIP	海域	东海	2009	黑潮贡献了 72% 的 DIN 输入和 84% 的 DIP 输入，57% 的 DIN 库存和 78% 的 DIP 库存	(Zhang et al., 2019)
	TP	湖泊	全球	1990—2010	磷引起的富营养化风险在发展中国家较高	(Fink et al., 2018)
	TP	次流域	美国	1945—2001	在美国，化肥和粪肥的使用量超过了作物利用率的 50%	(Sabo et al., 2021)
	DP	次流域县域	中国	2012	2012 年，中国河流中 75% 的污染来自点源，1/3 的子流域造成了一半以上的污染	(Chen et al., 2019)
	TP	0.1°经纬度次流域	长江流域	2017	2017 年长江流域 TP 输入： 点源—0.04 Mt 非点源—4.64 Mt	(Zou et al., 2020)
物质流清单分析	TP	省域	中国	2004	2004 年，中国磷的平均投入和产出分别为 28.9 kg/ha 和 14.2 kg/ha	(Chen et al., 2008)
	TP	省域	中国	2004	2004 年，中国农业向地表水排放磷总量 4.79×10^5 吨，其中畜牧业和矿肥排放的总磷影响最大	(Chen et al., 2010)

续表

类型	污染物	最小尺度	调查范围	调查时间段	结果	参考文献
物质流清单分析	TP	县域	中国	1990—2012	热点地区只占中国陆地面积的不到10%，却造成了超过一半的养分流入	(Wang et al., 2018)
	TP	国家级农业区	中国	1999—2019		(Chang and Zhang, 2023)
	TP	0.5°经纬度	全球	1900—2050	1950—2000年，全球总氮和总磷过剩量分别增加了7.76倍和44倍	(Bouwman et al., 2013)
	TP	5弧分	全球	2009—2012	Ecoinvent数据库低估了全球农业排放量的一个数量级	(Scherer and Pfister, 2015)
农业生产清单分析	TP	省域	中国	2009	中国海南、云南、广东和福建是世界上磷富营养化潜力最大的省份。湖北、湖南和四川三省氮、磷排放对当地富营养化潜力的贡献显著	(Huang et al., 2017)
	TP	省域	中国	1978—2017	1978—2017年，中国农业面源总磷污染增加了244.1%，其中畜禽养殖、农村生活垃圾和矿物肥对总磷的贡献率达69.1%~88.6%	(Zou et al., 2020)
城市排放清单分析	TP	WWTP	澳大利亚			(Foley et al., 2010)
	TP	WWTP	波兰	2011—2018	波兰污水处理厂产SSA的磷回收潜力达到1613.8 Mg	(Smol et al., 2020)
	TP	次流域	加拿大伊利湖	2014	化粪池排入伊利湖的P负荷占总负荷的1.7±0.8%~5±2.3%	(Oldfield et al., 2020)
	TP	养猪场	爱尔兰	2007		(Luo et al., 2014)
工业生产清单分析	TP	养猪场				
	TP	微藻生产设施	微藻生产设施		影响环境绩效的主要投入是电力使用、化肥需求（氮和磷）和运输	(Zhang et al., 2021)
	OPFRs	0.25°经纬度	中国	2014—2018	在OPFRs生产过程中造成的排放中，占比最高的行业是纺织（20.4%）、塑料（16.6%）和橡胶（4.3%）	(He et al., 2021)

然而，清单分析法在磷污染溯源领域也存在一定的局限性。首先，是数据的滞后性，根据表 4-1，有关清单分析法的文献发表时间要比他们调查的时间段晚数年，甚至数十年；其次，清单分析法难以基于磷形态进行溯源，在我们所调查的 27 篇关于清单分析法的文献中，有 21 篇对磷源的分析仅停留在 TP 层次；最后，清单分析法在实施过程中需要大量的污染源数据，且排放参数对解析结果的影响较大，导致结果不确定性大。

参考文献

[1] Bai Y, Sun C, Wang L, et al. 2022. The characteristics of net anthropogenic nitrogen and phosphorus inputs (NANI/NAPI) and TN/TP export fluxes in the Guangdong Section of the Pearl River (Zhujiang) Basin[J]. Sustainability 14(23), 16166.

[2] Bouwman L, Goldewijk K K, Van Der Hoek K W, et al. 2013. Exploring global changes in nitrogen and phosphorus cycles in agriculture induced by livestock production over the 1900-2050 period[J]. Proc. Natl. Acad. Sci. U. S. A. 110(52), 20882-20887.

[3] Chang D, Zhang Y. 2023. Farmland nutrient pollution and its evolutionary relationship with plantation economic development in China[J]. J. Environ. Manage. 325(Pt B), 116589.

[4] Chen M, Chen J, Sun F. 2008. Agricultural phosphorus flow and its environmental impacts in China[J]. Sci. Total Environ. 405(1-3), 140-152.

[5] Chen M, Chen J, Sun F. 2010. Estimating nutrient releases from agriculture in China: an extended substance flow analysis framework and a modeling tool[J]. Sci. Total Environ. 408(21), 5123-5136.

[6] Chen X, Strokal M, Van Vliet M T H, et al. 2019. Multi-scale modeling of nutrient pollution in the rivers of China[J]. Environ. Sci. Technol. 53(16), 9614-9625.

[7] Fink G, Alcamo J, Flörke M, et al. 2018. Phosphorus loadings to the world's largest lakes: Sources and trends[J]. Global Biogeochem. Cycles 32(4), 617-634.

[8] Foley J, de Haas D, Hartley K, et al. 2010. Comprehensive life cycle inventories of alternative wastewater treatment systems[J]. Water Res. 44(5), 1654-1666.

[9] He J, Wang Z, Zhao L, et al. 2021. Gridded emission inventory of organophosphorus flame retardants in China and inventory validation[J]. Environ. Pollut. 290, 118071.

[10] Herrera A, D'Imporzano G, Acién Fernandez F G, et al. 2021. Sustainable production of microalgae in raceways: Nutrients and water management as key factors influencing environmental impacts[J]. J. Clean. Prod. 287, 125005.

[11] Huang J, Xu C, Ridoutt B G, et al. 2017. Nitrogen and phosphorus losses and eutrophication potential associated with fertilizer application to cropland in China[J]. J. Clean. Prod. 159, 171-179.

[12] Li J C, Chen Y, Cai K K, et al. 2022. A high-resolution nutrient emission inventory for hotspot identification in the Yangtze River Basin[J]. J. Environ. Manage. 321, 115847.

[13] Li Y, Arocena J M, Zhang Q, et al. 2017. Heavy metals and nutrients (carbon, nitrogen, and phosphorus) in sediments: relationships to land uses, environmental risks, and management[J]. Environ. Sci. Pollut. Res. Int. 24(8), 7403-7412.

[14] Luo Y, Stichnothe H, Schuchardt F, et al. 2014. Life cycle assessment of manure management and nutrient recycling from a Chinese pig farm[J]. Waste Manag. Res. 32(1), 4-12.

[15] Oldfield L, Rakhimbekova S, Roy J W, et al. 2020. Estimation of phosphorus loads from septic systems to tributaries in the Canadian Lake Erie Basin[J]. J. Great Lakes Res. 46(6), 1559-1569.

[16] Pintado-Herrera M G, Wang C, Lu J, et al. 2017. Distribution, mass inventories, and ecological risk assessment of legacy and emerging contaminants in sediments from the Pearl River Estuary in China[J]. J. Hazard. Mater. 323(Pt A), 128-138.

[17] Sabo R D, Clark C M, Gibbs D A, et al. 2021. Phosphorus Inventory for the Conterminous United States (2002-2012) [J]. J. Geophys. Res. Biogeosci. 126(4), 1-21.

[18] Sabo R D, Sullivan B, Wu C, et al. 2022. Major point and nonpoint sources of nutrient pollution to surface water have declined throughout the Chesapeake Bay watershed[J]. Environ. Res. Commun. 4(4), 1-11.

[19] Scherer L, Pfister S. 2015. Modelling spatially explicit impacts from phosphorus emissions in agriculture. Int[J]. J. Life Cycle Assess. 20(6), 785-795.

[20] Smol M, Adam C, Anton Kugler S. 2020. Inventory of Polish municipal sewage sludge ash (SSA) - Mass flows, chemical composition, and phosphorus recovery potential[J]. Waste Manag. 116, 31-39.

[21] Strokal M, Kroeze C, Wang M, et al. 2016. The MARINA model (Model to Assess River Inputs of Nutrients to seAs): Model description and results for China[J]. Sci. Total Environ. 562, 869-888.

[22] Wang M, Ma L, Strokal M, et al. 2018. Hotspots for nitrogen and phosphorus losses from food production in China: A county-scale analysis[J]. Environ. Sci. Technol. 52(10), 5782-5791.

[23] Yang F, Wei Q, Chen H, et al. 2018. Long-term variations and influence factors of nutrients in the western North Yellow Sea, China[J]. Mar. Pollut. Bull. 135, 1026-1034.

[24] Zhang J, Guo X, Zhao L. 2019. Tracing external sources of nutrients in the East China Sea and evaluating their contributions to primary production[J]. Prog. Oceanogr. 176, 102122.

[25] Zhang Y, Jiang Y, Wang S, et al. 2021. Environmental sustainability assessment of pig manure mono- and co-digestion and dynamic land application of the digestate[J]. Renew. Sust. Energ. Rev. 137, 110476.

[26] Zou L, Liu Y, Wang Y, et al. 2020. Assessment and analysis of agricultural non-point source pollution loads in China: 1978-2017[J]. J. Environ. Manage. 263, 110400.

第 5 章

磷的源解析——扩散模型法

扩散模型通过模拟污染物从源到汇的水文过程和扩散到受体后对流域产生的影响实现源解析。在磷的源解析方面，主要应用的模型包括：基于物理的时间连续水文模型 SWAT；基于空间的非线性回归模型 SPARROW；基于暴雨事件聚焦农业面源污染的 AGNPS 模型；关注城市面源污染物的 SWMM 模型等。扩散模型可描述磷源对环境造成的影响，利用已知的影响因素（如来源个数和来源方位），识别主要污染源，估算污染源对受体的贡献率。本章将从模型的构建、应用和改进等方面介绍上述四种扩散模型。

5.1 SWAT 模型

5.1.1 SWAT 模型描述

SWAT（Soil and Water Assessment Tool）模型是一种基于物理，半分布式和时间连续的流域水文模型，广泛应用于模拟流域水文、土壤侵蚀、养分迁移和面源污染等过程（Ren et al., 2022a）。SWAT 是应用最广泛的水文模型，Fu et al.（2019）一篇综述统计了的 3282 篇文章，其中 44%使用了 SWAT 模型。

5.1.2 SWAT 模型构建与校准

利用 SWAT 模型对磷污染进行溯源过程中，通常需要空间数据[如高程模型（Digital elevation model，DEM）、地形数据、土地利用数据和土壤类型数据等]和属性数据（如气象数据、植被数据、水文数据和管理措施数据等）来准确模拟流域水文和水质响应过程。SWAT 模型构建过程如图 5-1 所示（彩图见附录 2）：① 利用空间数据统一坐标系，划分子流域；② 通过土地利用、土壤类型和坡面积最小值阈值划分水文响应单元（Hydrological response units，HRU）；③ 导入属性数据对模型进行模拟；④ 利用 SWATCUP 工具中的 Sequential Uncertainty Fitting（SUFI-2）算法，分析参数的敏感性和不确定性（Han et al., 2021; Hua et al., 2019; Ren et al., 2022a; Zuo et al., 2022）。

模型构建后，需要基于模拟值和观测值数据，利用相对误差（R_e）、模拟值和观测值的 Pearson 相关系数（r）、拟合度（R^2）、Nash-Sutcliffe 效率系数（Nash-Sutcliffe efficiency coefficient，E_{ns}）和改进 E_{ns} 系数（E'_{ns}）等评价参数对模型精度进行评价，具体公式如下：

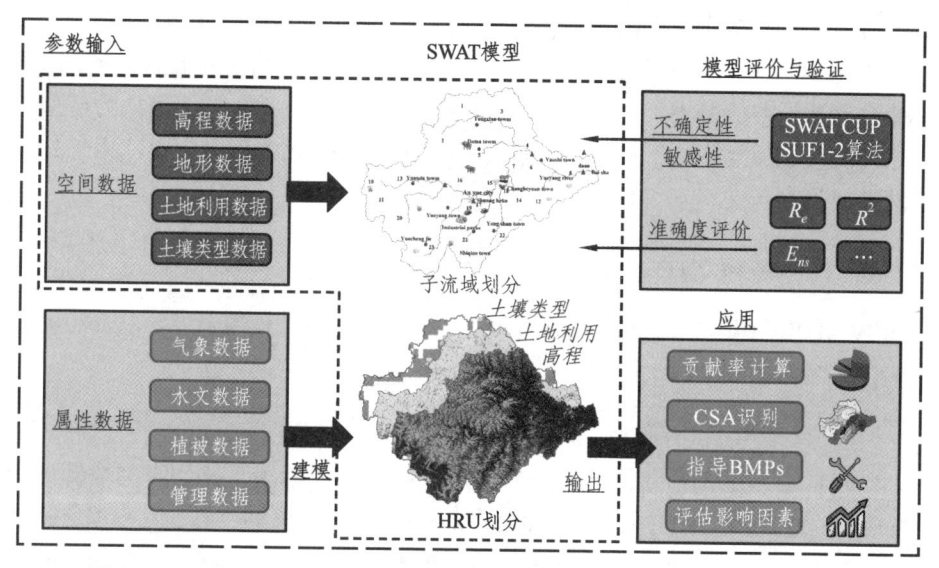

图 5-1 SWAT 建模过程，包括必要的参数输入、子流域和 HRU 划分、模型校正以及输出结果的应用

$$R_e = (\bar{S}-\bar{O})/\bar{O} \quad (5-1)$$

$$r = \frac{\sum_{i=1}^{n}(O_i-\bar{O})(S_i-\bar{S})}{\sqrt{\sum_{i=1}^{n}(O_i-\bar{O})^2}\sqrt{\sum_{i=1}^{n}(S_i-\bar{S})^2}} \quad (5-2)$$

$$R^2 = 1 - \frac{\sum_i(S_i-f_i)^2}{\sum_i(S_i-\bar{S})^2} \quad (5-3)$$

$$E_{ns} = 1 - \sum_{i=1}^{n}(O_i-S_i)^2/\sum_{i=1}^{n}(O_i-\bar{O})^2 \quad (5-4)$$

$$E'_{ns} = 1 - \sum_{i=1}^{n}|O_i-S_i|/\sum_{i=1}^{n}|O_i-\bar{O}| \quad (5-5)$$

式中 S_i 是模拟值，O_i 是观测值；$\bar{S}$ 和 $\bar{O}$ 分别是模拟值和观测值的均值；f_i 为模拟值和观测值之间的线性回归方程预测值（Li et al., 2015b）。

e 是相对误差，其绝对值越小表明模型模拟的性能越好；r 衡量的是线性相关关系，r 的绝对值越接近 1，线性相关性越强；R^2 是线性回归决定系数，表示模型解释的观测数据的变化，取值范围 0~1（1 表示观测数据的所有变化都由模型解释，通常认为大于 0.5 则模型是可信的）；E_{ns} 是一种归一化统计量，决定了观测数据（"信息"）中的方差与剩余方差（"噪声"）的相对大小，表示观测数据与预测数据在直线上的拟合程度，E_{ns}取值范围为$-\infty \sim 1$（1表示完全拟合，$0 \sim 1$是可以接受的性能级别）；相比于E_{ns}，E'_{ns}参数可以减少极值的影响，由于该值没有被平方值夸大，因此E'_{ns}值低于E_{ns}值，介于0.51和0.71可认为模型效率令人满意（Karki et al.，2017；Momm et al.，2022）。上述评价参数也适用于其他扩散模型，在后文中不再赘述。

5.1.3 SWAT模型的应用

SWAT模型可应用于磷源贡献率计算、CSAs识别、BMPs指导以及污染源影响因素的评估。

1. 磷源贡献率计算

SWAT最基本的应用之一是识别流域中磷的主要来源，并计算流域磷负荷（Cheng et al.，2021；Daloglu et al.，2012；Shen et al.，2014）。进一步地，根据SWAT的模拟结果，通常可以定量解析流域不同污染源的贡献率。例如，灯沙河流域和海河流域的主要磷源均为作物生产，贡献率分别为87.1%和73%（Han et al.，2021；Li et al.，2023）；Hua et al.（2019）利用养分输出系数联合SWAT量化了三峡库区湘西河流域养分负荷，结果表明湘西河流域农业源产生TP为97.4 t/a，其中畜牧业和农作物种植分别占52%和42%。Ren et al.（2022a）采用SWAT模型模拟了美国玉米种植带一个典型的暗沟排水面积流域，结果表明，化肥是磷输入的主要来源，占流域面积49%的玉米和大豆轮作系统，向流域贡献了88%的磷负荷。

2. 关键源区识别

SWAT可根据污染源定量解析结果识别流域内的关键源区（Critical Source Areas，CSAs）。CSAs是指在相对较短的时间内显著贡献更多的污染物区域（Chen et al.，2023；Shrestha et al.，2021）。流域不同区域的养分贡献水平可能存在很大差异，CSAs的识别是流域污染治理从全流域粗放型管理向精细化管理的前提（Chen et al.，2022a）。当前有不少研究利用SWAT识别了流域TP的CSAs（Han et al.，2021；Hua et al.，2019；Li et al.，2021；Ren et al.，2022a；Zuo et al.，2022），得到了不同尺度上的结果。例如，Li et al.（2021）在子流域尺度上识别了中国西北巴河流域面源污染空间分布特征并确定了CSAs；Zuo et al.（2022）和Shrestha et al.（2021）在HRUs尺度上分别识别了中国东北阿什河流域和加拿大安大略省Gully Creek流域TP的CSAs。

第 5 章 磷的源解析——扩散模型法

笔者在利用 SWAT 识别流域 CSAs 方面也有一定研究。发源于四川安岳县的岳阳河，在重庆市境内汇入琼江，是长江流域上游成渝地区重要的跨界小流域。该流域沟谷发达，稻田集中分布，主要的土地利用类型是耕地（87.50%）和林地（7.20%）。其中林地、房屋、个体户散养交织在一起，具备中国南方典型破碎景观格局（Flood et al., 2022）。同时由于跨界的地域属性，导致水体污染联防联治责任划分困难。使用 SWAT 模型识别该流域 CSAs，以求精细分析小流域农业面源污染来源贡献和时空变化，如图 5-2 所示（彩图见附录 2）。

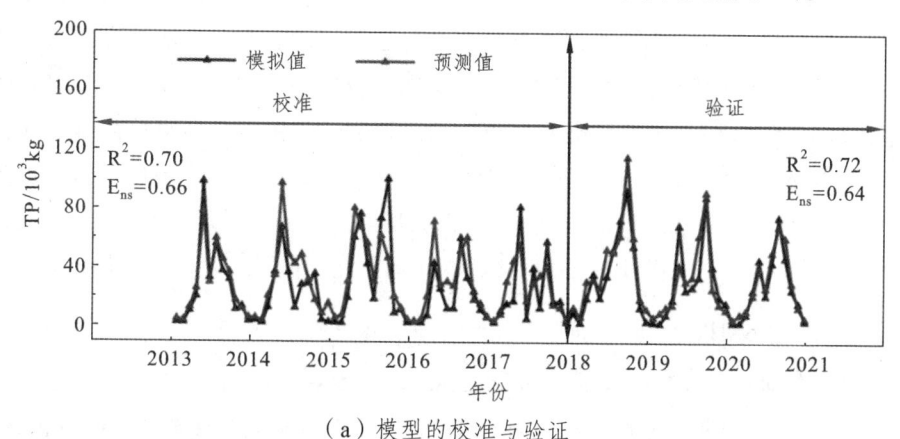

（a）模型的校准与验证

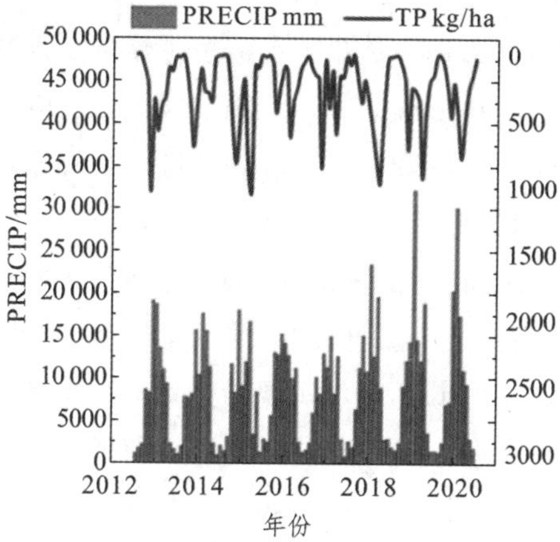

（b）时间步长 HRU 上的降雨总量与 TP 污染负荷数值关系

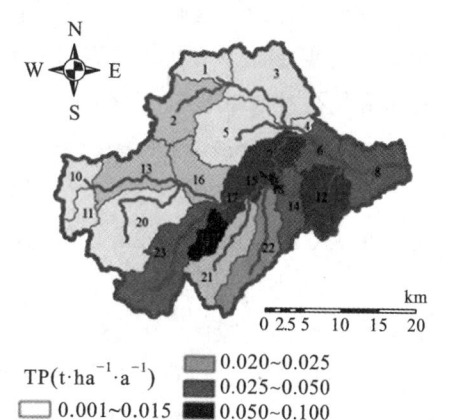

（c）次流域尺度的年平均氮磷负荷空间分布

图 5-2　(a) 模型的校准与验证；(b) 时间步长 HRU 上的降雨总量与 TP 污染负荷数值关系；(c) 次流域尺度的年平均氮磷负荷空间分布

如图 5-2 所示，首先基于模拟值和观测值对模型进行了校准和验证，在校准期和验证期，P 的 R^2 分别为 0.70、0.72，NSE 分别为 0.66、0.64，表明模型模拟氮磷的合理性较强，可用于模拟岳阳河流域农业面源污染。然后，分析了气候变化和农业面源污染的关系。岳阳河位于中国西南地区，夏秋季有着高温和高降水的特征。作物生长施用的大量化肥经降雨冲刷，导致地表径流中氮、磷浓度较高，随之迁移到水体中，加大了流域面源污染负荷（Yuan et al., 2022）。此外，研究区河道基本为自然坡岸，两岸的自然植被几乎被农业种植所取代，村庄建设用地和耕地也主要分布在河道沿线，农业及生活废弃物极易随着地表径流进入河流，极大地影响了水体质量（Zhang et al., 2016）。最终得到次流域尺度上年平均氮磷负荷空间分布。结果表明，岳阳河最高磷负荷出现在区域 19，达到 1.13 t·ha^{-1}·a^{-1}，其主要用地为建设用地和耕地；最低负荷出现在区域 5，为 0.011 t·ha^{-1}·a^{-1}，主要用地为林地和草地。流域平均磷负荷分别为 0.242 t·ha^{-1}·a^{-1}，城市用地的平均氮磷负荷最大，农田其次，林地最低。

3. 指导 BMPs 设置

最佳管理措施（Best Management Practices，BMPs）已被证明可以有效控制营养物质进入受纳水域（Chen et al., 2022a）。被污染流域的管理实践通常包含：保护性耕作、选择性施肥、退耕还林、梯田工程、人工湿地、草坡、作物轮作、肥料管理、温室雨水收集、等高线耕作、消除非法农业、水道植被恢复、海岸缓冲带和植被过滤带等形式（Chen et al., 2022a; Geng et al., 2019; Kast et al., 2021; Li et al., 2022a; Lopez-Ballesteros et al., 2023）。流域中大部分的面源污染通常由小部分子流域产生，SWAT 模型对 CSAs 的识别可指导 BMPs 的精准设置（Amin et al., 2017; Chen et al., 2022a; Uribe et al., 2018）。Li et al.（2022a）研究表明退耕还林适用于小范围内的面源污染控制，而对于大规模的面源污染控制，梯田工程是更有效的途径。此外，SWAT 指导 BMPs 设置，可实现污染负荷减量效果与经济成本的平衡，以最小化的设施维护成本，实现流域污染物负荷的最大削减（Geng et al., 2019; Liu et al., 2019）。

对于前文提到的岳阳河流域，笔者同样基于对 CSAs 识别，选择实施养分管理、土地利用管理和耕作管理三类 BMPs（Fenta et al., 2021），模拟其对研究区的实施效果，具体措施见表 5-1。

表 5-1 岳阳河流域 BMPs 设置以及措施代码

BMPs	措施	代码
基线		BAS
养分管理	化肥削减 10%	NM1
养分管理	化肥削减 20%	NM2
养分管理	免耕	NM3
土地利用管理	退耕还林	RF
土地利用管理	退耕育草	RG
土地利用管理	综合情景	RB
耕作管理	带状耕作	ST
耕作管理	作物残茬返田	RS
耕作管理	构建 5 m 河道缓冲带	CB

基于如表 5-1 的具体 BMPs 实施措施,得到流域尺度下不同 BMPs 的年均磷空间分布,如图 5-3 所示(彩图见附录 2)。

如图 5-3 所示,养分管理对污染物削减影响并不明显,减少化肥施用量使磷负荷平均削减 116.68 kg。土壤营养元素的含量受到长期耕作和施肥事件影响,短期的养分管理对解决面源污染的效果不太明显(Liu et al., 2016; Ren et al., 2021)。

实施土地利用管理措施,各种退耕情景下耕地的面积变化比例均为-23.1%。措施 RF、RG 和 RB 使得磷负荷分别削减了 609.04 kg、534.4 kg 和 602.04 kg。该结果表明,土地利用变化改变了面源污染空间负荷。林地、耕地、草地等表面覆盖物有着不同的属性(包括植被覆盖度、土壤根系结构、土壤属性等)(Keesstra et al., 2018)。不同准覆盖物会影响地表径流的产生,从而影响磷流失。土地利用管理在一定程度上可以降低面源污染的问题。

在耕作管理方面,措施 ST、RS 和 CB 使磷负荷分别削减了 926.4 kg、784.1 kg 和 1241.4 kg。ST 通过在小洼地中保留水分来减少地表径流,减少坡面和沟壑的侵蚀(Guto et al., 2011)。RS 措施能增加地表粗糙度、减缓地表径流,同时作物残茬覆盖也能使降雨更好入渗土壤,减少土壤水分蒸发(Tanveer et al., 2017)。CB 措施构建了植被缓冲带替代物理结构(Prosdocimi et al., 2016)。BMPs 研究结果表明,缓解研究区面源污染的最佳措施是实施耕作管理。实施耕作管理可以过滤地表径流、捕获沉积物、降低径流速度,使得降雨更好入渗土壤,恢复河岸缓冲带的功能,从而减少因土壤侵蚀进入水体的营养盐含量。

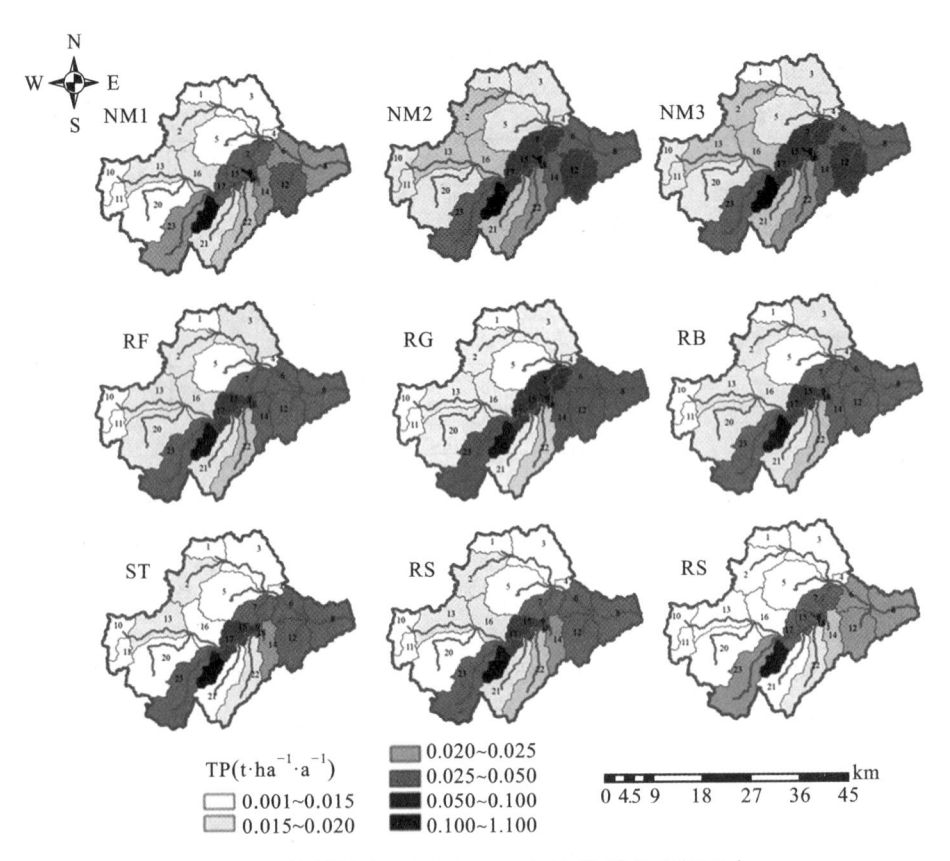

图 5-3 流域尺度下不同 BMPs 下年均磷的空间分布

4. 污染源影响因素评估

SWAT 模型侧重于评估土地利用、气候变化等因素对污染源的影响（El-Khoury et al., 2015; Ren et al., 2022b; Wang et al., 2018）。根据 SWAT 模拟结果，土地利用类型和土壤类型可显著影响 TP 负荷。在三峡库区，旱地和水田是养分负荷最大的土地利用类型，紫土、水田土和黄壤是养分负荷最大的土壤类型（Shen et al., 2013）。长期耕地扩张显著增加了磷负荷，而土壤性质的变化适度地抵消了这种影响（Huang et al., 2017）。

评估气候变化对磷污染的影响是当前 SWAT 的主要应用场景之一。Molina-Navarro et al.（2018）和 Mehan et al.（2019）通过 SWAT 评估了不同温室气体排放场景下的磷污染。他们的研究均表明，高排放负荷下，地表径流量将增大，导致 P_i、P_o 和 DP 等负荷上升（El-Khoury et al., 2015; Molina-Navarro et al., 2018）。

Li et al.（2022b）的 SWAT 模拟结果表明气候变化对 TP 负荷影响的权重达到了 90%，而土地演变对其影响具有累积效应。

5.1.4 SWAT 模型改进

SWAT 已被广泛应用于流域磷污染源解析，然而实际情况通常较为复杂。水质数据来源（Zhu et al., 2019）、DEM 分辨率、采样技术（Xu et al., 2016）、空间尺度（Chen et al., 2023）和气候变化（Wei et al., 2016）等因素均会带来不确定性，导致 SWAT 难以输出理想的结果。

此外，由于 SWAT 模型的半分布式和时间连续等特性，使得 SWAT 在特殊区域（如岩溶地区、山地低地混合流域、缺水流域和暗沟排水流域等）的应用将面对更大挑战（Amin et al., 2017; Izydorczyk et al., 2019; Pang et al., 2022; Zhang et al., 2022）。当前有大量的研究关注于 SWAT 模型的校正，以提高模型在这些特殊区域进行磷源解析的准确性。在山地低洼混合流域，由于水流方向不清晰，SWAT 模拟存在困难。Zhang et al.（2022）将 SWAT 与低地圩田磷动态模型拟合，可以很好地表征巢湖流域磷负荷（R^2>0.6，E_{ns}>0.6）；在缺水流域，连续性的 SWAT 模型会高估 TP 负荷。Pang et al.（2022）引入可靠性-弹性-脆弱性（R-R-V）指标构建的径流连续性指标与 SWAT 模型相结合，用于缺水流域 TP 负荷预测，R^2>0.80，E_{ns}>0.75；Bauwe et al.（2019）将排水和深层含水层中的 SRP 损失纳入 SWAT，对模型进行修正，用于暗沟排水流域，得到更满意结果（E_{ns}>0.5）。

5.2 SPARROW 模型

5.2.1 SPARROW 模型描述

SPARROW（Spatially Referenced Regression on Watershed Attributes）是一种基于空间的流域污染负荷计算非线性回归模型。使用质量平衡和统计方法来估计研究区域内某一组成部分的非保守运输（即带损失的运输），可以重现污染物从源头到目标出口的整个过程，揭示流域内起作用的重要环境因子（Robertson, Saad, 2021; Xu et al., 2021）。该模型使用混合统计和基于过程的方法在空间上将流域属性与河流联系起来，并提供了一个易于控制的框架，用于量化区域尺度上磷源和观测到的磷通量之间的空间显式关系（McLaughlin et al., 2022; Semadeni-Davies et al., 2020; Zhang et al., 2019）。目前，已应用于分析不同磷源及其贡献率（如施肥、牲畜饲料、大气源、磷矿、牧场源和林地等）。SPARROW还可以评估区域环境变量如景观变量、地下水补给、土壤成分、河流大小和水库池塘滞留等对磷源的影响（Xu et al., 2021）。SPARROW 模型的构建与应用过程如图 5-4 所示（彩图见附录 2）。

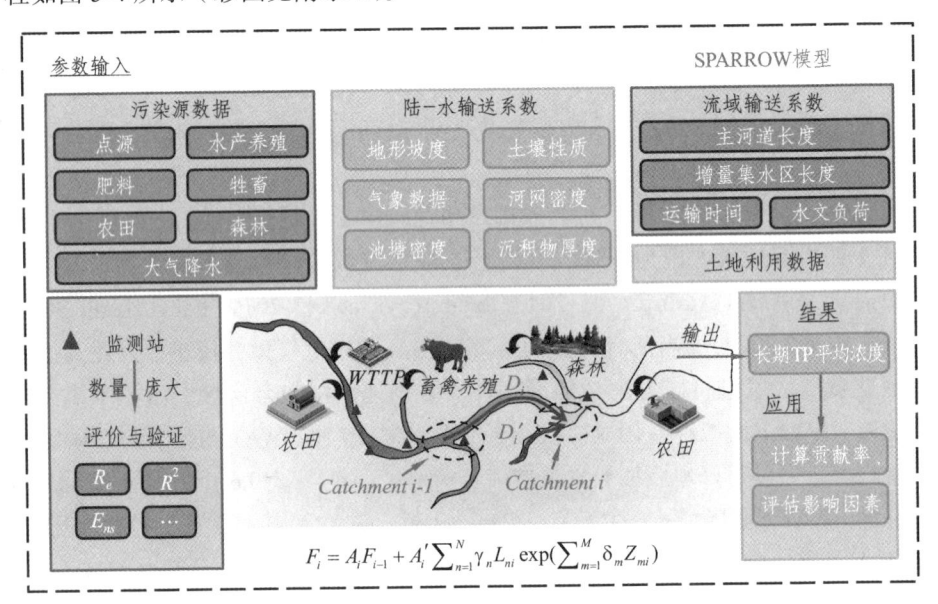

图 5-4 SPARROW 建模过程（包括必要的参数输入、模型校正以及输出结果的应用；除此之外图中还展示了干流增量、增量流域长度以及集水区之间的关系）

5.2.2 SPARROW 模型构建与校准

如图 5-4 所示，SPARROW 模型是对水文框架集水区进行的非线性最小二乘多元回归，以求解 TP 负荷的数学表达式，所需要的数据通常包括污染源数据（例如：点源输入、水产养殖、施肥、畜牧业、农田、大气沉降和林地等）、土地-水输送因子（Land-to-water delivery factor，如地形坡度、土壤性质、气温、水系密度、池塘密度和沉积物厚度等）、内流输送因子（In-stream delivery factor，如干流长度、增量集水区长度、输送时间和水力负荷等）以及土地利用面积（Garcia et al., 2016; Montefiore, Nelson, 2022; Xu et al., 2021; Zhang et al., 2019）。利用 SPARROW 对流域 TP 负荷进行源解析时，可使用式（5-6）计算：

$$F_i = A_i F_{i-1} + A'_i \sum_{n=1}^{N} \gamma_n L_{ni} \exp(\sum_{m=1}^{M} \delta_m Z_{mi}) \tag{5-6}$$

式中 F_i 为集水区 i 的 TP 年负荷，由两部分组成，A_iF_{i-1} 表示上游流域流入集水区 i 的 TP 负荷，$A_i'\sum_{n=1}^{N}\gamma_n L_{ni} \exp\left(\sum_{m=1}^{M}\delta_m Z_{mi}\right)$ 表示增量集水区内的 TP 负荷。在第一个求和项中，A_i 即为内流输送因子，表示 TP 通量沿河道从集水区 i-1 到 i 的衰减过程。当 i=1 时，F_{i-1}=0；在第二个求和项中，A_i' 表示增量集水区内的输送因子，$\gamma_n L_{ni}$ 表示源相（Source term），由源变量数组 L_{ni}（其中 n 取值范围从 1 到土地覆盖类型总数 N）和每种土地覆盖的源特定向量 γ_n 组成，该向量由自动参数估计获得。$\exp\left(\sum_{m=1}^{M}\delta_m Z_{mi}\right)$ 表示土地-水输送因子，由估计系数向量 δ_m 和流域属性数组 Z_{mi} 组成（其中 m 的取值范围从 1 到变量总数 M）（Xu et al., 2021; Zhang et al., 2019）。

A_i 和 A_i' 表示 TP 负荷在输送过程中的衰减，在 SPARROW 模型中假定二者遵循与传输距离相关的一级反应动力学，具体公式如下：

$$A_i = \exp(-\alpha \cdot D_i) \tag{5-7}$$

$$A_i' = \exp(-\alpha' \cdot D_i') \tag{5-8}$$

其中，D_i 表示将上游河段与集水区 i 连接起来的干流长度；D_i' 表示增量流域内的河流长度；α 和 α' 分别表示 TP 沿干流和支流随距离衰减的一阶衰减系数（图 5-4）。

收集数据后，利用流网络拓扑建立地理关系，使每个子流域和年份的每个输入数据分配一个唯一的水文序列代码（如在 SPARROW 数据文件中的变量"hydseq"）。以确保模型运行过程中能准确识别这些数据。模型构建后，需要利用观测点的数据对模型进行校准验证，这通常需要数量较大的观测点。在 Li et al.（2021），Zhou et al.（2018），Domagalski and Saleh（2015）和 Akomeah et al.（2021）的研究中分别使用了 27、30、80 和 85 个检测站的数据对模型进行校正。因此，实际中对检测站数量的要求可能会限制 SPARROW 模型的应用范围。

由于输入数据为年平均值，SPARROW 模型主要提供了长期平均养分负荷模式或个别基准年的非趋势负荷的空间分析，侧重于分析 TP 负荷的空间异质性。数据输出结果描述的是特定组分长期年均平均负荷（Neumann et al., 2023; Robertson and Saad, 2021），主要的应用场景为污染源解析和评估外界因素对污染源的影响。

5.2.3 SPARROW 模型的应用

1. 污染源贡献率计算

SPRAAOW 可定量分析 TP 来源，描绘具有较大营养输出倾向的"热点"，从而利于实施针对性管理（Ator et al., 2019; Li et al., 2015a; Morales-Marín et al., 2017; Neumann et al., 2023; Wang et al., 2021; Zhou et al., 2018）。例如，Zhang et al.（2019）使用 SPARROW 描述太湖流域上游农业快速扩张流域 TP 来源，表明农业、林地和城市用地分别占 TP 来源的 61%、21% 和 18%；Xu et al.（2021）利用 SPARROW 计算了天津市平原城市流域 2013 年的 TP 负荷，对上游排放、工业排放、污水排放、化肥施用、畜禽养殖、水产养殖和农村社区等磷源进行了定量解析。

2. 评估污染源影响因素

SPARROW 模拟结果侧重于体现磷源的空间异质性，可用于评估政策、人类活动和气候变化对磷源的影响。例如，McLaughlin et al.（2022）和 Garcia et al.（2016）利用 SPRAAOW 评估了 BMPs 对磷通量削减的影响以及农业保护措施和 TP 负荷之间的相关性关系；Wang et al.（2021）定量分析了人为扰动对黄河流域磷迁移的影响，表明农业活动是加剧流域磷污染的最主要因素；Akomeah et al.（2021）和 Neumann et al.（2023）评估排放情景和气候变化对 TP 负荷的影响，其中气候变化对 TP 负荷的影响实质上是通过影响地表径流而实现的。

通常，SPARROW 提供的模拟结果以年平均负荷的形式呈现，适用于估计长期稳态条件下 TP 在流域中的运动。该模型建立在基准年的基础上，得到的结果时间精度较粗略，且需要大量的监测点数据对模型进行校正（Domagalski, Saleh, 2015）。若要进一步发展 SPRAAOW 模型，需克服上述时间分辨率不足和监测点依赖性过大的问题。

5.2.4 SPARROW 模型改进

SPARROW 模型的本身属性决定了其在使用过程中存在一定的局限性（如时间分辨率不足和监测点依赖性过大）。SPARROW 最初被设计为一个静态建模框架，用于研究大型汇水区营养物质面源负荷的空间模式（McLaughlin et al., 2022; Wellen et al., 2012）。该模型难以捕捉流域营养负荷的季节性变化，也不能捕捉暴雨期间可能发生的负荷突变（Domagalski and Saleh, 2015）。Uribe et al.

（2018）和 Schmadel et al.（2021）试图将瞬时条件或季节性数据纳入模型，然而需要最大限度地增加监测站数量，以支撑模型的校准。Neumann et al.（2023）引入了具有信息参数先验的贝叶斯推理技术，提出了干/湿年分层贝叶斯框架（可纳入自适应流域的迭代检测、建模和评估，并根据新收集的支流水质数据进行顺序更新），以减少有限数量校准点与试图建模内容间的差距。并通过市政项目、学术发表、环保监测项目等数据来补充校准数据，以减少 SPARROW 模型对监测站数据的依赖。

此外，利用监测点对 SPARROW 模型进行校准的过程中，传统的校准技术假设所有的监测点的残差在统计上是独立的，并且计算监测负荷的误差相对较少。然而，监测距离较近且具有潜在较大误差的站点可能导致模型系数的精确度降低，从而使污染源的贡献率被过度分配到一个站点。Robertson and Saad（2021）开发了一种统计算法，根据上游流域面积比例计算每个残差的权重，用于加权非线性最小二乘法，对模拟系数进行重新优化。

5.3 AGNPS/AnnAGNPS 模型

5.3.1 AGNPS/AnnAGNPS 模型描述

AGNPS（Agricultural Non-Point Source Pollution Model）是一种基于事件的分布式模型，模拟单次暴雨事件产生的径流，被广泛应用于模拟不同流域水文、泥沙、农药和营养物质运移过程（León et al., 2004）。尽管 AGNPS 已发展为一个强大的流域模拟模型，然而在处理大量输入和输出数据时仍存在局限，且只适用于单一事件（Li et al., 2015b）。由此，研究者在 AGNPS 模型的基础上，将地理信息系统（Geographic Information System，GIS）数据处理技术与流域物理特征相结合，研发了 AnnAGNPS（Annualized Agricultural Non-Point Source Pollutant Model）模型（Li et al., 2015b）。这是一种流域尺度的连续模拟物理模型，为数据有限的地区提供了建模的可能（Karki et al., 2017; Li et al., 2015b）。当 AnnAGNPS 用于流域磷污染的源解析时，主要是根据进入和离开流域的磷的总量，通过质量平衡对降雨事件中磷的负荷进行估计（Li et al., 2015b）。模型模拟结果可用于指导和评估 BMPs 的效果（Momm et al., 2022）。AGNPS/AnnAGNPS 模型的构建与校准过程如图 5-5 所示（彩图见附录 2）。

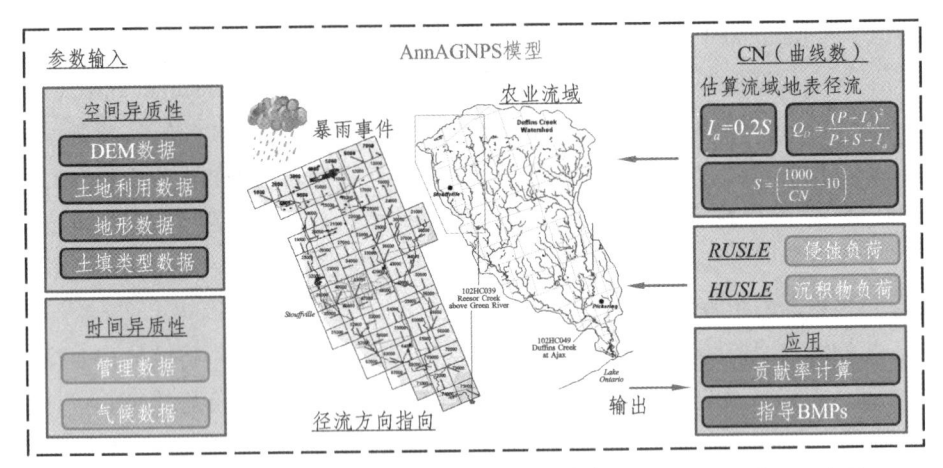

图 5-5　AGNPS/AnnAGNPS 模型的构建[图中参考了 León et al.（2004）对 AGNPS 模型的描述，以示意该模型是如何划分工作区并通过 DEM 数据模拟暴雨事件中的径流方向]

5.3.2　AGNPS/AnnAGNPS 模型构建与校准

AGNPS 是一种分布式模型，模拟单一暴雨事件的农业流域径流，并假设降水是均匀的。该模型建模的流域必须划分为均匀的方形工作区，称之为单元（Cell）（León et al., 2004）。而单元则通过被称为河段（Reach）的通道相互连接（Que et al., 2015）。AnnAGNPS 模型是 AGNPS 的延续，具有批处理、连续模拟、每日时间步长和流域尺度等标签。旨在模拟径流、五种不同粒径沉积物（如黏土、粉砂、砂土、小团聚体和大团聚体）、农药和营养物质（如氮、磷和有机物）在流域中的运输（Karki et al., 2017）。由于 AnnAGNPS 模型是 AGNPS 模型的延续，该节着重描述 AnnAGNPS 模型的建模过程。

AnnAGNPS 模型为农业流域设计，可以使用 33 种数据输入集（如流域数据、沟渠数据、点源数据、蓄水数据、化肥施用数、农药施用数据……）。该模型建立过程中必要的参数包括流域物理特征数据（DEM 数据和土壤数据）、土地利用数据、管理操作数据和气候数据。其中流域物理特征数据和土地利用数据构成了模型的空间变异性，而管理操作数据和气候数据构成了模型的时间变异性（Karki et al., 2017）（图 5-5）。

通过 TOPAGNPS（一种基于 GIS 的景观分析组件），利用 DEM 数据确定流域空间特征，将流域划分为均匀的方形单元和流经河段的路线（图 5-4）（Karki et al., 2017; Li et al., 2015b）。使用土壤保持服务（Soil Conservation Service，SCS）

曲线数（Curve Number，CN）法估计流域地表径流，进而计算河流径流量和地下径流量。SCS 是分配给土地利用、土壤和先前土壤湿度条件（Antecedent soil Moisture Condition，AMC）的独特离散组合的任意数，是流域预测模型中最重要的参数。计算过程见以下公式：

$$I_a = 0.2S \tag{5-9}$$

$$S = \left(\frac{1000}{CN} - 10\right) \tag{5-10}$$

$$Q_D = \frac{(P - I_a)^2}{P + S - I_a} \tag{5-11}$$

式中 CN 是曲线数，可以在模型中调整，以考虑整个流域土地利用变化（Karki et al., 2017）；Q_D 是地表径流量（Surface runoff），mm；P 是暴雨降雨量（Storm precipitation），mm；S 是潜在最大滞留量（Potential maximum retention），对于暴雨事件，S 是恒定的，mm；I_a 是初始情景（Initial abstraction）考虑了地表径流形成前的入渗、截留、洼地储存和 AMC 过程，mm（Gupta et al., 2019）。

AnnAGNPS 模型采用"修正通用土壤流失方程法"（Revised Universal Soil Loss Equation, RUSLE）估计坡面侵蚀和细沟侵蚀，采用"水文地貌通用土壤流失方程法"（Hydro-geomorphic Universal Soil Loss Equation, HUSLE）预测暴雨期间的沉积负荷（Sediment load）（Karki et al., 2017）。与其他扩散模型一样，AnnAGNPS 模型构建后需要对其进行校准和验证。利用 AnnAGNPS 进行磷负荷模拟时，通常需要多个月平均数据。例如 Li et al.（2015）等人利用 AnnAGNPS 模型对太湖流域进行养分负荷模拟时，利用了 2008 年 7 月至 2009 年 9 月的观测数据对模型进行校准，然后使用 2012 年 12 月至 2013 年 12 月的养分数据进行验证（Li et al., 2015b）。评价校准和验证效果的参数包括 R^2 和 E_{ns} 等。模拟结果可用于计算磷污染负荷，BMPs 指导和污染源影响因素评估。

5.3.3 AGNPS/AnnAGNPS 模型的应用

1. 污染负荷计算与 BMPs

AGNPS 和 AnnAGNPS 均可用于农业面源磷的污染负荷估算（Kao et al., 1998; Zhu et al., 2015），模拟结果可指导管理者设置河岸缓冲林、减肥、免耕、平行梯田和控制排水（Controlled tile drainage）等常规 BMPs（Chen et al., 2022b;

Gupta et al., 2019; Jiang et al., 2019; Que et al., 2015）。沟渠通常在重大降水事件后反复出现，对污染物输送起到关键作用，常规 BMPs 很难专注于沟渠控制（Gupta et al., 2019）。基于事件的 AGNPS 模型可以很好地指导如何设置 Water and Sediment Control Basin（WASCoB，一种专门用于阻碍沟渠发展的 BMP）（Gupta et al., 2019）。该 BMP 通过在河流斜坡上建造蓄水池、护堤、进水口和排水沟，旨在捕获水流并缓慢排放，从而针对性地对流域内反复出现的沟渠进行污染物输送的有效控制。

2. 评估污染源影响因素

AGNPS/AnnAGNPS 模型兼具时间异质性和空间异质性，侧重于评估 BMPs（如河岸缓冲林、减肥、免耕、平行梯田和控制排水等）（Chen et al., 2022b; Jiang et al., 2019; Que et al., 2015）和其他人类活动或管理措施对面源磷污染的影响（Zhu et al., 2020）。例如，Karki et al.（2017）使用 AnnAGNPS 评估了密西西比河东部农业流域以农场蓄水（OFWS）作为 BMPs 的实施效果，结果表明，模型中磷的月负荷估算 R^2=0.74，E_{ns}=0.54，OFWS 截留了 558 kg 磷，防止了下游的养分污染；Jiang et al.（2019）使用 AnnAGNPS 评估了不同宽度河岸缓冲林对总磷负荷的影响，结果表明 20 m、40 m 和 60 m 河岸缓冲林使流域 TP 负荷分别减少了 18.16%、45.93%和 56.2%，结合径流和其他营养物质的削减效果，最终确定 40 m 宽度为最佳控制宽度；Xu et al.（2012）使用 AnnAGNPS 评估了土地利用变化对非点源磷污染的影响，结果表明，快速城市化会导致非点源污染加剧。

5.3.4 AGNPS/AnnAGNPS 模型改进

AGNPS 是基于单一事件的分布式模型，AnnAGNPS 模型作为 AGNPS 的改进型已得到了广泛应用。同样地，由于 AnnAGNPS 模型本身的局限性，该模型还有很多值得改进的方面。Li et al.（2015b）评估了 AnnAGNPS 在小流域径流和氮磷负荷估算的性能和适用性，结果表明，磷负荷模拟的校准和验证结果稍差，R^2 分别为 0.60 和 0.83，E_{ns} 分别为 0.61 和-3.86。然而，目前尚未发现 AnnAGNPS 针对小流域改进的相关报道。

另外，AnnAGNPS 在建模过程中只考虑了降雨的径流响应，未考虑其对受纳水体的影响。Chao et al.（2023）将考虑泥沙和水质组分相互作用的 CCHE-WQ 模型与 AnnAGNPS 相结合。由于 CCHE-WQ 模型需要入口处的流量和营养物质浓度，将 AnnAGNPS 模拟结果作为 CCHE-WQ 模型的边界条件，使得两种模型在功能上实现了互补。耦合后的模型用于研究湖泊水质对农业管理措施的响应，考虑了泥沙与水质组分之间的相互作用，可以系统地模拟流域和径流、泥沙和养分过程。

5.4 SWMM 模型

5.4.1 SWMM 模型描述

SWMM（Storm Water Management Model）模型是一种基于动态的分布式水文水力模型，因其操作简单，源代码开放，计算效率高，广泛用于城市雨水管理和水质模拟（Ma et al., 2022; Nayeb Yazdi et al., 2019）。城市化的过程中，将增加地面的不透水性，通过渠化、管道和自然排水通道改变流域水文，增加了径流量，减少了入渗，导致向下游输送更多养分（Nayeb Yazdi et al., 2019）。SWMM 将城市下垫面划分为若干子集水区，并通过节点和管网组合起来。每个子集水区被概括为透水区和不透水区，水流路径基于质量、能量和动量守恒方程。可用于模拟离散事件和连续时期的径流量和水质，评价城市发展的影响，调查流域恢复策略的有效性（Ma et al., 2022; Nayeb Yazdi et al., 2019; Nayeb Yazdi et al., 2021）。

城市雨水污染物负荷本质上是一种面源污染，其特征是污染物种类繁多，来源各异（Tuomela et al., 2019）。SWMM 提供了一个水质包（Water Quality Package）以实现城市径流水质模拟（Tuomela et al., 2019）。在 SWMM 中，子集水区可以连接到下游的子集水区、下水道系统，并创建流动路径。子集水区接收来自降雨和上游相连的子集水区径流和污染物，在计算如洼地储存、入渗和蒸发等损失后，得到向下游子集水区产生的径流和污染物负荷（图 5-6，彩图见附录 2）。通常，SWMM 可同时对 TN、氨氮（NH_4^+-N）、总悬浮物（TSS）、COD 和 TP 等多种污染物模拟和污染负荷进行估算（Taghizadeh et al., 2021; Tang et al., 2021）。并指导雨水控制措施，如低影响发展（Low Impact Development, LID）、绿色雨水基础设施（Green Stormwater Infrastructure, GSI）和海绵城市建设（Sponge City Construction），预测和评价城市面源污染控制的 BMPs 功效（Dai et al., 2020; Nayeb Yazdi et al., 2019）。

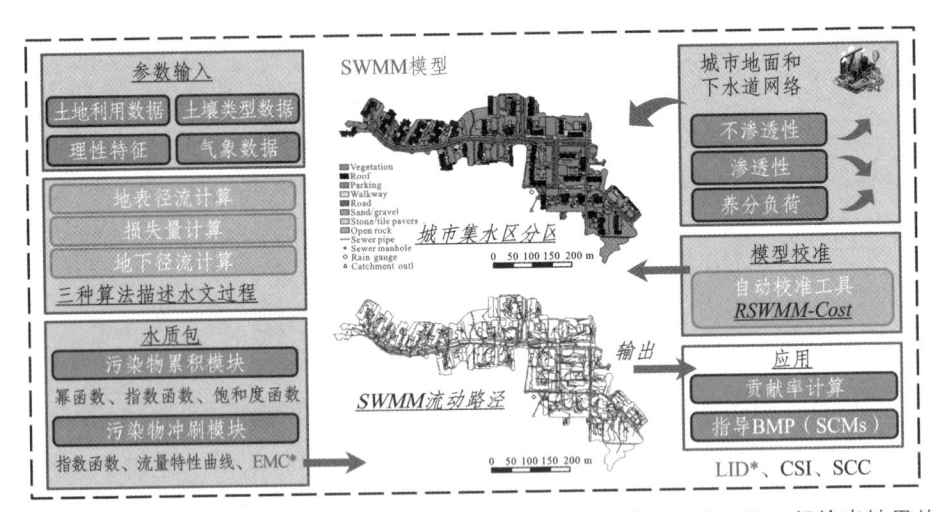

图 5-6 SWMM 建模过程[包括必要的参数输入、水质包、模型校准工具已经输出结果的应用；图中参考了 Tuomela et al.（2019）关于 SWMM 模型的描述，以示意该模型对于城市子集水区的划分以及模拟的径流和管网流动路径]

5.4.2 SWMM 模型构建与校准

在 SWMM 模型中，子集水区是基本水文单元。该模型集成了三组算法来描述暴雨期间发生的水文过程，包括地表径流算法、损失算法和地下流量算法（Nayeb Yazdi et al., 2019）。其中，地表径流算法是基于 DEM 和土地覆盖信息，计算了流域面积、平均地表坡度、地表流道宽度和不透水率 4 个物理特征；损失算法用于估算蒸发和入渗造成的雨水损失，通常根据气象资料确定蒸发量，基于 Horton 法计算入渗量；地下流量算法用于模拟管道中的水流，通常由动态波路由法实现（图 5-6）（Dai et al., 2020; Wang et al., 2023）。

SWMM 提供了水质模块，使其可以将功能扩展到城市面源污染评估。对于每次模拟，模型将自动生成子集水区摘要（Subcatchment summaries）和来自上游地区的贡献。具体步骤为：① 为了确定每个子集水区的污染物冲刷量，移除上游相连的子集水区的径流输入，并根据剩余的径流量计算负荷；② 根据土地覆盖类型对子流域进行分组，计算总源区负荷和径流量；③ 通过对源区负荷和流量进行求和，计算流域总负荷和流域总径流量；④ 通过将总流域负荷除以模拟总径流量，计算出每个模拟时段总流域面积的平均体积加权平均浓度。在后面的结果中，这被称为站点平均浓度（Site mean concentration, SMC）（Tuomela et al., 2019）。

在 SWMM 的水质模块中，需计算污染物的累积量与冲刷量，这是 SWMM 模拟水质过程中最重要的两个参数。污染物的累积采用指数函数、幂函数或饱和度方程来模拟；而污染物的冲刷通常采用指数函数、评级曲线方程和事件平均浓度法（Event Mean Concentration，EMC）来估算（Tuomela et al., 2019），这部分计算过程可参考 SWMM 手册（Rossman and Huber, 2016）。具体计算过程和公式如下。

1. 污染物累积函数

（1）幂函数，污染物的累积与时间成正比，直到达到最大极限。

$$b = \text{Min}(B_{\max}, K_B t^{N_B}) \tag{5-12}$$

b 为污染物累积；t 为累积时间间隔；$B_{\max}$ 为可能的最大累计量；K_B 为累积速率常数；N_B 为累积时间指数，应≤1，以便随时间增加累积速率减小，当 $N_B = 1$ 时，得到一个线性累积函数。

（2）指数函数，污染物累积遵循指数增长，渐进地接近最大极限。

$$b = B_{\max}(1 - e^{-K_B t}) \tag{5-13}$$

（3）饱和度方程，以线性速率开始，随着时间不断推移，逐渐下降，直到达到饱和。

$$b = B_{\max} t / (K_B + t) \tag{5-14}$$

该公式中 K_B 为 1/2 饱和常数，即达到最大累计一般的天数。

2. 污染物冲刷函数

（1）指数冲刷。

$$w = K_w q^{N_w} m_B \tag{5-15}$$

式中，w 冲刷速率；K_w 为冲刷系数；q^{N_w} 为校正后的子集水区径流速率，详细校正过程见 SWMM 手册；m_B 为表面污染物质量。

（2）评级曲线方程。

$$w = K_w Q^{N_w} \tag{5-16}$$

$$Q = q f_{LU} A \tag{5-17}$$

式中，Q 为流速；K_w 和 Q^{N_w} 为冲刷系数；$f_{LU}A$ 是指在总集水区面积 A 中用于被分析的比例。

（3）事件平均浓度。

$$w = K_w q f_{LU} A \tag{5-18}$$

式中，K_w 为 EMC。

SWMM 参考手册提供了各种累积和冲刷方程的比较，以供选择最合适的方程（Rossman and Huber，2016）。在模拟暴雨水质和估算污染物负荷时，从业者更倾向于使用 EMCs。这是由于其他方程对数据有较高要求，涉及的参数没有现场数据很难校准。而不同污染物、源区域和地理位置的 EMCs 更容易获得。为了获得可行的校准和验证并控制计算负担，使用 EMCs 简化模型已被证明是有效的（Tuomela et al.，2019）。对于城市磷面源污染负荷的计算，使用 EMC 评估城市径流中的 TP 淋滤过程，即通过输入不同土地利用类型的 TP 的 EMC 来估算其负荷（Furtado et al.，2021）。

模型建立后，SWMM 使用自动校准工具 RSWMM-Cost，调整所有 19 个特定模型参数（e.g. 水力宽度、水力传导性、坡度和不透水性），使每个子集水区都在指定范围内（Alamdari et al.，2022）。模拟值和观测值仍通过 R^2 和 E_{ns} 等参数进行评估。

5.4.3　SWMM 模型的应用

1. SWMM 污染负荷计算

SWMM 提供的水质包可用于估算城市面源的磷负荷，倾向于在较小尺度（2 ha ~ 4000 km^2）上捕捉暴雨事件的响应（Barone et al.，2019；Furtado et al.，2021；Nayeb Yazdi et al.，2019；Yazdi et al.，2019）。例如，Barone et al.（2019）研究了意大利一深湖流域下水道溢流对湖泊的贡献，SWMM 估计下水道系统中 22%的磷进入了该湖泊。Alamdari et al.（2022）关注暴雨对地表累积污染物的初期冲刷效应，SWMM 模拟结果表明，超过 50%的 TP 被最初的 20%径流转移。此外，SWMM 也可用于评估城市流域气候和土地利用变化对污染物负荷的影响，气候变化会导致 TP 负荷季节性波动，土地利用变化则主要是通过影响城市地面的不透水性，从而改变城市面源的磷负荷。

2. BMPs 设置以及效果评估

SWMM 模型污染负荷估算的详细结果可指导城市管理者制定合适的 BMPs

策略,以控制磷的城市面源污染。雨水控制措施作为城市特定的 BMPs,以储存的方式来缓解峰值径流量,从而降低对城市地表水体的影响(Nayeb Yazdi et al.,2019)。SWMM 可通过模拟径流的运输过程来指导城市 BMPs 设置。LID 是应用最广泛的城市 BMPs,多项研究均提到了 SWMM 对 LID 设置的指导与优化(Baek et al., 2020; Hashemi and Mahjouri, 2022; Taghizadeh et al., 2021; Yang et al., 2023)。同时,SWMM 亦可用于评估不同 BMPs 对面源污染的控制效果(Baek et al., 2020; Nayeb Yazdi et al., 2021; Phillips et al., 2022; Tang et al., 2021)。Nayeb Yazdi et al.(2021)评估了滞留塘对雨水养分的控制效果,结果表明滞留塘可使流出的 TP 减少 10%。Phillips et al.(2022)评估了河流绿色设施对集水区规模养分的干预,表明在 6 种不同情景下 TP 去除率达到 59%~90%。

5.4.4 SWMM 模型改进与不确定性

由于 SWMM 本身的特性,模型仍存在各方面的不足,根据模型本身的特性对其进行改进,有助于 SWMM 模型在城市面源污染解析方面的推广应用。例如,SWMM 建模过程中通常有数百个或数十个子集水区,需要大量的场地参数来充分表达降雨和径流之间的复杂关系,这使得模型的校正具有挑战。Shahed Behrouz et al.(2020)将 SWMM 与 OSTRICH(一种与模型无关的优化校准工具)相结合,最大限度减少了峰值流量和总流量误差之间的相互冲突。

此外,SWMM 模拟结果存在一定的不确定性,这些不确定性来源包括,城市高度空间异质性使得模型需要高分辨率下垫面数据,而这样的数据往往难以获得;人工建筑、不透水路面和道路排水管道改变了径流自然路径,使得水力交换条件复杂;对离散分布的子集水区溢流过程进行仿真,仿真结果可靠性难以衡量;SWMM 模型通过 EMC 估算污染物负荷,然而基于文献得来的 EMC 值往往不符合当地条件;EMC 法没有反应暴雨事件中水量变化对污染物浓度稀释的影响(Ma et al., 2022; Tuomela et al., 2019)。Dai et al.(2020)利用 Cellular automate 算法替换了原本的 SWMM 陆上路由模块,一定程度上解决了市政排水数据缺乏的问题。不过,其他方面的不确定性该如何改进亟待研究。

5.5 扩散模型综合评价

除上述提及的 SWAT、SPARROW、AGNPS 和 SWMM 模型外,还有 Hydrologic Simulation Program-Fortran (HSPF), Pollutant Loading Estimator

（PLOAD）和 Distributed Hydrology Soil Vegetation Model（DHSVM）等扩散模型也在磷污染解析方面有所应用（Bah et al., 2020）。这些模型具有各自的特点，在复杂性、数据要求以及最终提供的输出等方面存在差异，因此不存在适用于所有应用场景的"最佳模型"。选择模型时应遵循"为正确的理由选择正确的模型"的原则。考虑到研究最终目标和需要产出的尺度，在选择一种模式之前需要仔细审查所需数据的广度和质量，模型的复杂性以及流域的物理特征（Abdelwahab et al., 2018）。

另外，为了推广扩散模型在更多场景下的应用，针对模型本身的特点进行改进是当前研究的热点与难点（表 5-2）。不过，扩散模型的复杂性和对密集数据要求是其应用拓展最大的挑战。

表 5-2 磷的源解析-扩散模型改进综合概述

模型	存在问题	污染物	尺度	改进方式	结果	参考文献
SWAT	喀斯特地貌使水文运输复杂化	TP	子流域	使用 Topo-SWAT 工具	校准和验证 E_{ns}=0.79 或 0.73	(Amin et al., 2017)
	不同 DEM 分辨率对模型灵敏度有影响	TP	子流域			(Zhu et al., 2019)
	山地与低地混合流域水流方向不确定	TP	子流域	低地圩田系统磷耦合动态模型	R^2>0.6, E_{ns}>0.6	(Zhang et al., 2022)
	SWAT 会高估缺水流域的总磷	TP	子流域	结合可靠性-弹性-脆弱性构建的水流连续性指标	R^2>0.8, E_{ns}>0.75	(Pang et al., 2022)
	没有考虑暗沟排水流域的 DRP	DRP	子集水区	DRP 损失被纳入 SWAT	E_{ns}>0.5	(Bauwe et al., 2019)
	畜禽粪便排放对 SWAT 结果的影响	P_o P_i	子流域	创建新的粪便数据库	P_o: R^2=0.77, E_{ns}=0.65 P_i: R^2=0.75, E_{ns}=0.61	(Liu et al., 2017)
	气候模式不能完全代表观测数据	DRP	子流域	偏差纠正	提高了精度,但影响了养分负荷和水文过程变化的方向和幅度	(Miralha et al., 2021)
SPARROW	暴雨事件期间,很难捕捉养分负荷和负荷的季节变化	TP	集水区	引入瞬时条件和季节数据	需要最大限度地增加监测站的数量	(Domagalski and Saleh, 2015)

续表

模型	存在问题	污染物	尺度	改进方式	结果	参考文献
SPARROW	难以捕捉养分负荷的季节变化和过度依赖监测站数据	TP	子流域	提出干湿年份的分层贝叶斯框架	减少对监测站数据的依赖	(Neumann et al., 2023)
SPARROW	少数距离过近的监测点对结果影响偏大	TP	集水区	根据上游流域面积为每个监测点分配权重		(Robertson and Saad, 2021)
AGNPS	小流域磷负荷模拟结果较差	TP	子流域			(Li et al., 2015b)
AnnAGNPS	未考虑污染物对水体的影响	PO_4^-	流域	耦合 CCHE-WQ 模型	两种模式实现了功能互补	(Chao et al., 2023)
SWMM	模型校准对参数要求高		城市子集水区	耦合 OSTRICH 工具	最大流量与总流量误差之间的冲突最小	(Shahed Behrouz et al., 2020)
SWMM	难以获得高分辨率的下垫面数据		城市子集水区			(Ma et al., 2022)
SWMM	市政排水数据难以获得		城市子集水区	Cellular automate 算法取代 SWMM 陆路模块		(Dai et al., 2020)
SWMM	恒定的 EMC 带来了模型不确定性		城市子集水区			(Tuomela et al., 2019)

参考文献

[1] Abdelwahab O M M, Ricci G F, De Girolamo, et al. 2018. Modelling soil erosion in a Mediterranean watershed: Comparison between SWAT and AnnAGNPS models[J]. Environ. Res. 166, 363-376.

[2] Akomeah E, Morales-Marin L A, Carr M, et al. 2021. The impacts of changing climate and streamflow on nutrient speciation in a large Prairie reservoir[J]. J. Environ. Manage. 288, 112262.

[3] Alamdari N, Claggett P, Sample D J, et al. 2022. Evaluating the joint effects of climate and land use change on runoff and pollutant loading in a rapidly developing watershed[J]. J. Clean. Prod. 330, 129953.

[4] Amin M G M, Veith T L, Collick A S, et al. 2017. Simulating hydrological and nonpoint source pollution processes in a karst watershed: A variable source area hydrology model evaluation[J]. Agric. Water Manag. 180, 212-223.

[5] Ator S W, García A M, Schwarz G E, et al. 2019. Toward explaining nitrogen and phosphorus trends in Chesapeake Bay Tributaries, 1992-2012[J]. J. Am. Water Resour. Assoc. 55(5), 1149-1168.

[6] Baek S S, Ligaray M, Pyo J, et al. 2020. A novel water quality module of the SWMM model for assessing low impact development (LID) in urban watersheds[J]. J. Hydrol. 586, 124886.

[7] Bah H, Zhou M, Ren X, et al. 2020. Effects of organic amendment applications on nitrogen and phosphorus losses from sloping cropland in the upper Yangtze River[J]. Agric. Ecosyst. Environ. 302, 107086.

[8] Barone L, Pilotti M, Valerio G, et al. 2019. Analysis of the residual nutrient load from a combined sewer system in a watershed of a deep Italian lake[J]. J. Hydrol. 571, 202-213.

[9] Bauwe A, Eckhardt K U, Lennartz B. 2019. Predicting dissolved reactive phosphorus in tile-drained catchments using a modified SWAT model[J]. Ecohydrol. Hydrobiol. 19(2), 198-209.

[10] Chao X, Witthaus L, Bingner R, et al. 2023. An integrated watershed and water quality modeling system to study lake water quality responses to agricultural

management practices[J]. Environ. Monit. Assess. 164, 105691.

[11] Chen L, Li J, Xu J, et al. 2022a. New framework for nonpoint source pollution management based on downscaling priority management areas[J]. J. Hydrol. 606, 127433.

[12] Chen L, Xu Y, Li S, et al. 2023. New method for scaling nonpoint source pollution by integrating the SWAT model and IHA-based indicators[J]. J. Environ. Manage. 325(Pt A), 116491.

[13] Chen Y, Lu B, Xu C, et al. 2022b. Uncertainty Evaluation of Best Management Practice Effectiveness Based on the AnnAGNPS Model[J]. Water Resour. Manag. 36(4), 1307-1321.

[14] Cheng J, Gong Y, Zhu D Z, et al. 2021. Modeling the sources and retention of phosphorus nutrient in a coastal river system in China using SWAT[J]. J. Environ. Manage. 278(Pt 2), 111556.

[15] Dai Y, Chen L, Shen Z. 2020. A cellular automata (CA)-based method to improve the SWMM performance with scarce drainage data and its spatial scale effect[J]. J. Hydrol. 581, 124402.

[16] Daloglu I, Cho K H, Scavia D. 2012. Evaluating causes of trends in long-term dissolved reactive phosphorus loads to Lake Erie[J]. Environ. Sci. Technol. 46(19), 10660-10666.

[17] Domagalski J, Saleh D. 2015. Sources and transport of phosphorus to rivers in California and Adjacent States, U.S., as determined by SPARROW modeling[J]. J. Am. Water Resour. Assoc. 51(6), 1463-1486.

[18] El-Khoury A, Seidou O, Lapen DR, et al. 2015. Combined impacts of future climate and land use changes on discharge, nitrogen and phosphorus loads for a Canadian river basin[J]. J. Environ. Manage. 151, 76-86.

[19] Fenta A A, Tsunekawa A, Haregeweyn N, et al. 2021. Agroecology-based soil erosion assessment for better conservation planning in Ethiopian river basins[J]. Environ. Res. 195, 110786.

[20] Flood M T, Hernandez-Suarez J S, Nejadhashemi A P, et al. 2022. Connecting microbial, nutrient, physiochemical, and land use variables for the evaluation of water quality within mixed use watersheds[J]. Water Res. 219, 118526.

[21] Fu B H, Merritt W S, Croke B F W, et al. 2019. A review of catchment-scale water quality and erosion models and a synthesis of future prospects[J]. Environ. Model Softw. 114, 75-97.

[22] Furtado A, Monte-Mor R C A, Couto E A D. 2021. Evaluation of reduction of external load of total phosphorus and total suspended solids for rehabilitation of urban lakes[J]. J. Environ. Manage. 296, 113339.

[23] Garcia A M, Alexander R B, Arnold J G, et al. 2016. Regional effects of agricultural conservation practices on nutrient transport in the upper Mississippi River Basin. Environ[J]. Sci. Technol. 50(13), 6991-7000.

[24] Geng R, Yin P, Sharpley A N. 2019. A coupled model system to optimize the best management practices for nonpoint source pollution control[J]. J. Clean. Prod. 220, 581-592.

[25] Gupta A K, Rudra R P, Gharabaghi B, et al. 2019. CoBAGNPS: A toolbox for simulating water and sediment control basin, WASCoB through AGNPS model[J]. Catena 179, 49-65.

[26] Guto S N, Pypers P, Vanlauwe B, et al. 2011. Tillage and vegetative barrier effects on soil conservation and short-term economic benefits in the Central Kenya highlands[J]. Field Crops Res. 122(2), 85-94.

[27] Han J, Xin Z, Han F, et al. 2021. Source contribution analysis of nutrient pollution in a P-rich watershed: Implications for integrated water quality management[J]. Environ. Pollut. 279, 116885.

[28] Hashemi M, Mahjouri N. 2022. Global Sensitivity Analysis-based Design of Low Impact Development Practices for Urban Runoff Management Under Uncertainty[J]. Water Resour. Manag. 36(9), 2953-2972.

[29] Hua L, Li W, Zhai L, et al. 2019. An innovative approach to identifying agricultural pollution sources and loads by using nutrient export coefficients in watershed modeling[J]. J. Hydrol. 571, 322-331.

[30] Huang H, Ouyang W, Wu H, et al. 2017. Long-term diffuse phosphorus pollution dynamics under the combined influence of land use and soil property variations[J]. Sci. Total Environ. 579, 1894-1903.

[31] Izydorczyk K, Piniewski M, Krauze K, et al. 2019. The ecohydrological approach, SWAT modelling, and multi-stakeholder engagement - A system

solution to diffuse pollution in the Pilica basin, Poland[J]. J. Environ. Manage. 248, 109329.

[32] Jiang K, Li Z, Luo C, et al. 2019. The reduction effects of riparian reforestation on runoff and nutrient export based on AnnAGNPS model in a small typical watershed, China[J]. Environ. Sci. Pollut. Res. Int. 26(6), 5934-5943.

[33] Kao J J, Lin W L, Tsai C H. 1998. Dynamic spatial modeling approach for estimation of internal phosphorus load[J]. Water Res. 32(1), 47-56.

[34] Karki R, Tagert M L M, Paz J O, et al. 2017. Application of AnnAGNPS to model an agricultural watershed in East-Central Mississippi for the evaluation of an on-farm water storage (OFWS) system[J]. Agric. Water Manag. 192, 103-114.

[35] Kast J B, Kalcic M, Wilson R, et al. 2021. Evaluating the efficacy of targeting options for conservation practice adoption on watershed-scale phosphorus reductions[J]. Water Res. 201, 117375.

[36] Keesstra S, Nunes J, Novara A, et al. 2018. The superior effect of nature based solutions in land management for enhancing ecosystem services[J]. Sci. Total Environ. 610, 997-1009.

[37] León L F, Booty W G, Bowen G S, et al. 2004. Validation of an agricultural non-point source model in a watershed in southern Ontario[J]. Agric. Water Manag. 65(1), 59-75.

[38] Li H, Zhou X, Huang K, Hao G, et al. 2022a. Research on optimal control of non-point source pollution: a case study from the Danjiang River basin in China[J]. Environ. Sci. Pollut. Res. Int. 29(11), 15582-15602.

[39] Li S, Li J, Xia J, et al. 2021. Optimal control of nonpoint source pollution in the Bahe River Basin, Northwest China, based on the SWAT model. Environ[J]. Sci. Pollut. Res. Int. 28(39), 55330-55343.

[40] Li X, Wellen C, Liu G, et al. 2015a. Estimation of nutrient sources and transport using Spatially Referenced Regressions on Watershed Attributes: a case study in Songhuajiang River Basin, China[J]. Environ. Sci. Pollut. Res. Int. 22(9), 6989-7001.

[41] Li X, Xu W, Song S, et al. 2023. Sources and spatiotemporal distribution

characteristics of nitrogen and phosphorus loads in the Haihe River Basin, China[J]. Mar. Pollut. Bull. 189, 114756.

[42] Li Y, Wang H, Deng Y, et al. 2022b. How climate change and land-use evolution relates to the non-point source pollution in a typical watershed of China[J]. Sci. Total Environ. 839, 156375.

[43] Li Z, Luo C, Xi Q, et al. 2015b. Assessment of the AnnAGNPS model in simulating runoff and nutrients in a typical small watershed in the Taihu Lake basin, China[J]. Catena 133, 349-361.

[44] Liu R, Wang Q, Xu F, et al. 2017. Impacts of manure application on SWAT model outputs in the Xiangxi River watershed[J]. J. Hydrol. 555, 479-488.

[45] Liu X, Sheng H, Jiang S Y, et al. 2016. Intensification of phosphorus cycling in China since the 1600 s. Proc. Natl. Acad. Sci. U. S. A. 113(10), 2609-2614.

[46] Liu Y, Guo T, Wang R, et al. 2019. A SWAT-based optimization tool for obtaining cost-effective strategies for agricultural conservation practice implementation at watershed scales[J]. Sci. Total Environ. 691, 685-696.

[47] Lopez-Ballesteros A, Trolle D, Srinivasan R, et al. 2023. Assessing the effectiveness of potential best management practices for science-informed decision support at the watershed scale: The case of the Mar Menor coastal lagoon, Spain[J]. Sci. Total Environ. 859(Pt 1), 160144.

[48] Ma B, Wu Z, Hu C, et al. 2022. Process-oriented SWMM real-time correction and urban flood dynamic simulation[J]. J. Hydrol. 605, 127269.

[49] McLaughlin P, Alexander R, Blomquist J, et al. 2022. Power analysis for detecting the effects of best management practices on reducing nitrogen and phosphorus fluxes to the Chesapeake Bay Watershed, USA[J]. Ecol. Indic. 136, 108713.

[50] Miralha L, Muenich R L, Scavia D, et al. 2021. Bias correction of climate model outputs influences watershed model nutrient load predictions[J]. Sci. Total Environ. 759, 143039.

[51] Molina-Navarro E, Andersen H E, Nielsen A, et al. 2018. Quantifying the combined effects of land use and climate changes on stream flow and nutrient loads: A modelling approach in the Odense Fjord catchment (Denmark)[J]. Sci. Total Environ. 621, 253-264.

[52] Momm H G, Bingner R L, Moore K, et al. 2022. Integrated surface and groundwater modeling to enhance water resource sustainability in agricultural watersheds[J]. Agric. Water Manag. 269, 107692.

[53] Montefiore L R, Nelson N G. 2022. Can a simple water quality model effectively estimate runoff-driven nutrient loads to estuarine systems? A national-scale comparison of STEPLgrid and SPARROW[J]. Environ. Model Softw. 150, 105344.

[54] Morales-Marín L A, Wheater H S, Lindenschmidt K E. 2017. Assessment of nutrient loadings of a large multipurpose prairie reservoir[J]. J. Hydrol. 550, 166-185.

[55] Nayeb Yazdi M, Ketabchy M, Sample D J, et al. 2019. An evaluation of HSPF and SWMM for simulating streamflow regimes in an urban watershed[J]. Environ. Monit. Assess. 118, 211-225.

[56] Nayeb Yazdi M, Scott D, Sample D J, et al. 2021. Efficacy of a retention pond in treating stormwater nutrients and sediment[J]. J. Clean. Prod. 290, 125787.

[57] Neumann A, Blukacz-Richards E A, Saha R, et al. 2023. A Bayesian hierarchical spatially explicit modelling framework to examine phosphorus export between contrasting flow regimes[J]. J. Great Lakes Res. 49(1), 190-208.

[58] Pang S, Wang X, Melching C S, et al. 2022. Identification of multilevel priority management areas for diffuse pollutants based on streamflow continuity in a water-deficient watershed[J]. J. Clean. Prod. 351, 131322.

[59] Phillips D, Jamwal P, Lindquist M, et al. 2022. Assessing catchment-scale performance of in-stream Provisional Green Infrastructure interventions for Dry Weather Flows[J]. Landsc. Urban Plan. 226, 104448.

[60] Prosdocimi M, Tarolli P, Cerdà A. 2016. Mulching practices for reducing soil water erosion: A review[J]. Earth Sci. Rev. 161, 191-203.

[61] Que Z, Seidou O, Droste R L, et al. 2015. Using AnnAGNPS to Predict the Effects of Tile Drainage Control on Nutrient and Sediment Loads for a River Basin[J]. J. Environ. Qual. 44(2), 629-641.

[62] Ren C C, Jin S Q, Wu Y Y, et al. 2021. Fertilizer overuse in Chinese smallholders due to lack of fixed inputs[J]. J. Environ. Manage. 293, 112913.

[63] Ren D, Engel B, Mercado J A V, et al. 2022a. Modeling and assessing water and nutrient balances in a tile-drained agricultural watershed in the U.S. Corn Belt[J]. Water Res. 210, 117976.

[64] Ren D, Engel B, Tuinstra M R. 2022b. Crop improvement influences on water quantity and quality processes in an agricultural watershed[J]. Water Res. 217, 118353.

[65] Robertson D M, Saad D A. 2021. Nitrogen and phosphorus sources and delivery from the Mississippi/Atchafalaya River Basin: An update using 2012 SPARROW models[J]. J. Am. Water Resour. Assoc. 57(3), 406-429.

[66] Rossman L A, Huber W C. 2016. Storm water management model reference manual volume III–Water quality, US EPA National Risk Management Laboratory, Cincinnati, Ohio.

[67] Schmadel N M, Harvey J W, Schwarz G E. 2021. Seasonally dynamic nutrient modeling quantifies storage lags and time-varying reactivity across large river basins[J]. Environ. Res. Lett. 16(9).

[68] Semadeni-Davies, A, Jones-Todd C, Srinivasan M S, et al. 2020. CLUES model calibration and its implications for estimating contaminant attenuation[J]. Agric. Water Manag. 228.

[69] Shahed Behrouz M, Zhu Z, Matott L S, et al. 2020. A new tool for automatic calibration of the Storm Water Management Model (SWMM)[J]. J. Hydrol. 581, 124436.

[70] Shen Z, Chen L, Hong Q, et al. 2013. Assessment of nitrogen and phosphorus loads and causal factors from different land use and soil types in the Three Gorges Reservoir Area[J]. Sci. Total Environ. 454-455, 383-392.

[71] Shen Z, Qiu J, Hong Q, et al. 2014. Simulation of spatial and temporal distributions of non-point source pollution load in the Three Gorges Reservoir Region. Sci. Total Environ. 493, 138-146.

[72] Shrestha N K, Rudra R P, Daggupati P, et al. 2021. A comparative evaluation of the continuous and event-based modelling approaches for identifying critical source areas for sediment and phosphorus losses[J]. J. Environ. Manage. 277, 111427.

[73] Taghizadeh S, Khani S, Rajaee T. 2021. Hybrid SWMM and particle swarm optimization model for urban runoff water quality control by using green infrastructures (LID-BMPs)[J]. Urban For. Urban Green. 60.

[74] Tang S, Jiang J, Zheng Y, et al. 2021. Robustness analysis of storm water quality modelling with LID infrastructures from natural event-based field monitoring[J]. Sci. Total Environ. 753, 142007.

[75] Tanveer M, Anjum S A, Hussain S, et al. 2017. Relay cropping as a sustainable approach: problems and opportunities for sustainable crop production[J]. Environ. Sci. Pollut. Res. Int. 24(8), 6973-6988.

[76] Tuomela C, Sillanpaa N, Koivusalo H. 2019. Assessment of stormwater pollutant loads and source area contributions with storm water management model (SWMM)[J]. J. Environ. Manage. 233, 719-727.

[77] Uribe N, Corzo G, Quintero M, et al. 2018. Impact of conservation tillage on nitrogen and phosphorus runoff losses in a potato crop system in Fuquene watershed, Colombia[J]. Agric. Water Manag. 209, 62-72.

[78] Wang L, Hou H, Li Y, et al. 2023. Investigating relationships between landscape patterns and surface runoff from a spatial distribution and intensity perspective[J]. J. Environ. Manage. 325(Pt B), 116631.

[79] Wang Q, Liu R, Men C, et al. 2018. Application of genetic algorithm to land use optimization for non-point source pollution control based on CLUE-S and SWAT[J]. J. Hydrol. 560, 86-96.

[80] Wang Y, Ouyang W, Zhang Y, et al. 2021. Quantify phosphorus transport distinction of different reaches to estuary under long-term anthropogenic perturbation[J]. Sci. Total Environ. 780, 146647.

[81] Wei P, Ouyang W, Hao F, et al. 2016. Combined impacts of precipitation and temperature on diffuse phosphorus pollution loading and critical source area identification in a freeze-thaw area[J]. Sci. Total Environ. 553, 607-616.

[82] Wellen C, Arhonditsis G B, Labencki T, et al. 2012. A Bayesian methodological framework for accommodating interannual variability of nutrient loading with the SPARROW model[J]. Water Resour. Res. 48(10), W10505.

[83] Xu F, Dong G, Wang Q, et al. 2016. Impacts of DEM uncertainties on critical source areas identification for non-point source pollution control based on SWAT model[J]. J. Hydrol. 540, 355-367.

[84] Xu K, Wang Y, Su H, et al. 2012. Effect of land-use changes on nonpoint source pollution in the Xizhi River watershed, Guangdong, China. Hydrol[J]. Process. 27(18), 2557-2566.

[85] Xu Z, Ji Z, Liang B, et al. 2021. Estimate of nutrient sources and transport into Bohai Bay in China from a lower plain urban watershed using a SPARROW model[J]. Environ. Sci. Pollut. Res. Int. 28(20), 25733-25747.

[86] Yang B, Zhang T, Li J, et al. 2023. Optimal designs of LID based on LID experiments and SWMM for a small-scale community in Tianjin, north China[J]. J. Environ. Manage. 334, 117442.

[87] Yazdi M N, Sample D J, Scott D, et al. 2019. Water quality characterization of storm and irrigation runoff from a container nursery[J]. Sci. Total. Environ. 667, 166-178.

[88] Yuan L, Fang L C, Wang Y Z, et al. 2022. Anthropogenic activities accelerated the evolution of river trophic status. Ecol. Indic. 136, 108584.

[89] Zhang J, Gao J, Zhu Q, et al. 2022. Coupling mountain and lowland watershed models to characterize nutrient loading: An eight-year investigation in Lake Chaohu Basin[J]. J. Hydrol. 612, 128258.

[90] Zhang W, Pueppke S G, Li H, et al. 2019. Modeling phosphorus sources and transport in a headwater catchment with rapid agricultural expansion[J]. Environ. Pollut. 255(Pt 2), 113273.

[91] Zhang Y L, Shi K, Liu J J, et al. 2016. Meteorological and hydrological conditions driving the formation and disappearance of black blooms, an ecological disaster phenomena of eutrophication and algal blooms[J]. Sci. Total Environ. 569, 1517-1529.

[92] Zhou P, Huang J, Hong H. 2018. Modeling nutrient sources, transport and management strategies in a coastal watershed, Southeast China[J]. Sci. Total Environ. 610-611, 1298-1309.

[93] Zhu K W, Chen Y C, Zhang S, et al. 2020. Output risk evolution analysis of

agricultural non-point source pollution under different scenarios based on multi-model[J]. Glob. Ecol. Conserv. 23.

[94] Zhu Q D, Sun J H, Hua G F, et al. 2015. Runoff characteristics and non-point source pollution analysis in the Taihu Lake Basin: a case study of the town of Xueyan, China[J]. Environ. Sci. Pollut. Res. Int. 22(19), 15029-15036.

[95] Zhu Y, Chen L, Wei G, Li S, et al. 2019. Uncertainty assessment in baseflow nonpoint source pollution prediction: The impacts of hydrographic separation methods, data sources and baseflow period assumptions[J]. J. Hydrol. 574, 915-925.

[96] Zuo D, Han Y, Gao X, et al. 2022. Identification of priority management areas for non-point source pollution based on critical source areas in an agricultural watershed of Northeast China[J]. Environ. Res. 214(Pt 2), 113892.

第 6 章

磷的源解析——受体模型法

受体模型法是通过测量流域内不同位置受体的理化性质，定性地识别受体中的污染源，定量地计算污染源对受体的贡献率。该方法在进行源解析时，不需要污染源的排放条件、周围环境等数据，也不用追踪污染物的迁移与转化规律，规避了对复杂参数的需求。在磷污染源解析方面，应用较广的受体模型法有多元统计法和稳定同位素示踪法。

6.1 多元统计分析

多元统计分析法运用数学工具，通过耦合污染源成分与河流水体成分，建立污染指标与其来源的因果关系与贡献大小（Zhang et al., 2020a; Zhang et al., 2020b）。该方法通过受体检测获得的已知信息之间的相互关系，形成源成分谱或能够表示污染源类型的污染因子，由此反向揭示污染来源并确定其贡献率，极大地减小了源解析的工作量（Zhang et al., 2020a）。常用的模型有 PCA/FA、APCS-MLR 和 PMF 模型，后文将从模型的描述、建立和应用等方面依次介绍这几种模型。

6.1.1 PCA/FA

PCA（Principal Component Analysis）通过将相关性较强的多个变量降维至少数几个主成分，来反映数据的主要变化模式（Zhang et al., 2020b）。在这个过程中，原始变量通过线性变换组合成为新的主成分，主成分的数量比原始变量少，但它们能够解释原始数据中大部分的变异（Zhang et al., 2020b）；FA（Factor Analysis）从复杂数据集中提取共性因子并解释变量之间的相关性。在这个过程中，原始变量被假设由少数几个共性因子所解释，每个变量可以被解释为各因子对其的贡献之和，每个因子又可以由多个变量解释（Fu et al., 2020）。PCA 和 FA 表达式分别见式（6-1）和式（6-2）

$$Z_{ij} = a_{i1}x_{1j} + a_{i2}x_{2j} + \cdots + a_{in}x_{nj} \tag{6-1}$$

$$Z_{ij} = f_{1i}\partial_{f1} + f_{2i}\partial_{f2} + \cdots + f_{mi}\partial_{fm} + e_{fi} \tag{6-2}$$

式中，Z 为成分得分；a 为成分负荷；x 为变量实测值；i 为组分数；j 为样本数；n 为变量总数。

PCA/FA 可以将复杂数据集中潜在的相关变量线性组合，然后通过最大方差旋转将主成分上变量负荷极化，形成新的主成分，成为变因子（VFs）。变量负荷在 0.3~0.5、0.5~0.75、0.75~1 范围内分别定义为低、中、强负荷，同一 VF 上具有强负荷的多个变量通常表明他们受同一源的影响（Chen et al., 2022）。该模型可以定性地研究水污染中的潜在污染源，但污染源对受体的定量贡献不能由 PCA/FA 直接得到。

此外，PCA/FA 的数据集需要足够的自由度（如样本数—变量数＞50），并且还需通过 Kaiser-Meyer-Olkin（KMO）和 Bartlett's 球度检验（p）对变量之间的简单相关系数进行评估：KMO＞0.5 和 p＜0.05 表明数据集适合 PCA/FA（Chen et al., 2022; Zhang et al., 2020b）。

6.1.2 APCS-MLR

APCS-MLR 是一种利用多元线性回归（Multiple Linear Regression, MLR）和自适应加权化学平衡模型（Adaptive Potential-Concentration Surface, APCS）相结合的统计模型。可以通过对受体样品的化学组成进行多元线性回归分析，确定不同污染源贡献的比例，并进一步利用 APCS 模型来优化回归结果（Zhang et al., 2020a）。在 APCS 分析前必须进行 PCA 分析以获得潜在污染源的定性信息（Xie et al., 2023）。

具体地，APCS-MLR 建模步骤为：① 根据受体采样数据，采用变差旋转的 PCA 提取特征值＞1 的成分，识别污染源；② 基于 PCA 得分分析，确定各样本的绝对主成分评分（APCS）；③ 对每种样本的 APCS 与样品质量浓度进行多元线性回归，得到基本 MLR 方程；④ 利用 MLR 模型估计每个样本的源贡献（Zhang et al., 2018）。APCS-MLR 假设所有潜在污染源对受体位置的特定污染物最终浓度有线性贡献（Xie et al., 2023）。

在上述步骤②中，由于数据集已经归一化，故 PCA 生成的样本因子得分不能直接用于计算因子的贡献。因为这些结果是基于 Z 转化（Z-transformed）的标准化变量，需要将归一化得分重新缩放转换为未归一化的 APCS 对应值（Xie et al., 2023）。

APCS-MLR 通过构建多个线性回归方程，减去零值浓度样本的因子得分，可以量化每个源对变量的实际贡献。基于不同源的线性组合，某系数的观测浓度计算公式如下（Xie et al., 2023）：

$$C_i = (\gamma_0)_i + \sum_{j=1}^{p} \gamma_{ji} \times APCS_{ji} \qquad (6-3)$$

式中，C_i 为参数 i 的观测浓度；$(\gamma_0)_i$ 为方程常数项，表示未知源的贡献；γ_{ji} 为第 j 个源对第 i 个变量的回归系数；$APCS_{ji}$ 表示考虑样本旋转的缩放值；$\gamma_{ji} \times APCS_{ji}$ 表示第 j 个源对第 i 个变量浓度的贡献（Xie et al., 2023）。

由于多元线性回归计算中可能出现的负系数会导致对源的负贡献，从而混淆数据解释。可以将负贡献转化为绝对值，来重新分配源贡献。

$$S_i = \frac{abs(S_i) \times 100\%}{abs(UN) + \sum_{j=1}^{p} abs(S_j)} \tag{6-4}$$

式中，S_i 是第 i 个来源的贡献；UN 是位置来源的贡献；p 为确定来源的总数（Zhang et al., 2022）。

6.1.3　PMF

PMF（Positive Matrix Factorization）是一种基于因子分析的受体模型，将浓度矩阵分解为因子贡献矩阵和因子分布矩阵。最初是由美国环保署开发，用于解析大气颗粒中的污染源。该模型的核心原则是使用最小二乘法确定每个源因子的分布和贡献，并对源贡献应用非负约束（Ren et al., 2023b）。具体地，通过多次迭代最小二乘法最小化模型的残差和，以获得 PMF 的最佳因子贡献曲线。

$$x_{ij} = \sum_{k=1}^{p} g_{ki} f_{kj} + e_{ij} \tag{6-5}$$

式中，x_{ij} 表示第 i 个样品第 j 个水化学参数的观测值；g_{ki} 为第 k 个源对第 i 个样品的贡献；f_{kj} 为第 k 个源中，第 j 个水化学参数的浓度；p 为污染源总数；e_{ij} 为误差。

在 PMF 中以信噪比（S/N）作为数据质量的指标，S/N＞1 表示为数据质量正常；0.5~1 之间表示数据质量较弱，需要对所有样本进行三倍不确定加权；S/N＜0.2 表示数据质量差（Ren et al., 2023b）。

除上述提到的 PCA/FA、PMF 和 APCS-MLR 模型外，回归分析、聚类分析（Cluster analysis, CA）、模糊综合评价（Fuzzy Comprehensive Assessment, FCA）和 UNMIX 等模型也常用于污染物源解析。其中 UNMIX 模型在 PM10、PM2.5、VOCs 和重金属等污染物源解析方面有较为广泛的应用，然而与磷源解析相关报道还不多见。

6.1.4 多元统计应用与综合评述

PCA/FA、APCS-MLR 和 PMF 等多元统计模型被广泛用于流域污染源解析，与前文提到的扩散模型不同的是，多元统计法以更宏观更全面的时间关注流域的整体污染情况，而不是某种特定的污染物。对于流域磷污染，多元统计模型会将该因素考虑其中，但是很少单独关注磷这一种特定的污染物。例如，Zhang et al.（2020）基于 2017—2018 年 15 种水质参数数据（包括 TP），利用 APCS-MLR 模型对岷江的污染源进行了解析。结果表明在重度污染地区，工业废水和生活污染、城市污水、农业面源、浮游植物生长以及自然和季节性影响对流域污染贡献分别为 20.81%、16.66%、15.73%、12.86%和 11.59%（Zhang et al., 2022）。

多元统计模型用于流域磷污染综合评价见图 6-1。在多元统计模型当中，磷通常作为一个污染参数参与贡献率的计算。多元统计模型通常相互配合，能够识别 3~6 个污染源，以表示流域的整体污染情况，TP 是常用的参数，而其他磷形态往往很少参与多元统计模型计算。

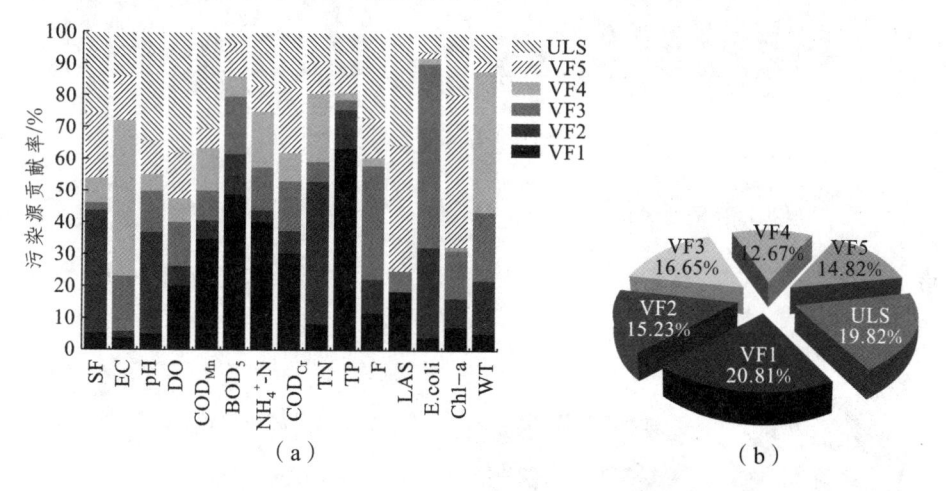

图 6-1　基于污染源解析解释的不同污染源对水质参数的贡献（a）和对流域的评价贡献（b）（Zhang et al., 2022）

对于各多元统计模型，PCA 模型可以提取隐藏在大型复杂数据集中的重要信息，建立起水质变量与污染源之间的关系。然而，PCA 难以获得污染源的定量信息，存在一定的运用局限性（Xie et al., 2023）；PMF 是一种先进的因子分析技术，其每个数据点都是不确定性加权的，而不像 PCA 那样，每个数据点权重相同（Qi et al., 2020）。相比于 PMF，APCS-MLR 模型对数据中的异常值不敏感,且通过正交变换得到的源谱更容易识别,性能更稳定(Xie et al., 2023)。Zhang et al.(2020)对比了 PMF 和 APCS-MLR 两种受体模型在四川盆地混合土地利用地区的适用性(图 6-2)。其结果表明 PMF 和 APCS-MLR 两种模型均确定了 5 种地下水污染源,两种模型的平均源贡献在农业源上存在显著差异。与 PCA-APCS-MLR 模型相比,PMF 模型中较高的 R^2 和较小的未解释度,表明 PMF 方法可以在四川盆地混合土地利用地区提供更合理的来源分配,更真实地反映地下水污染。

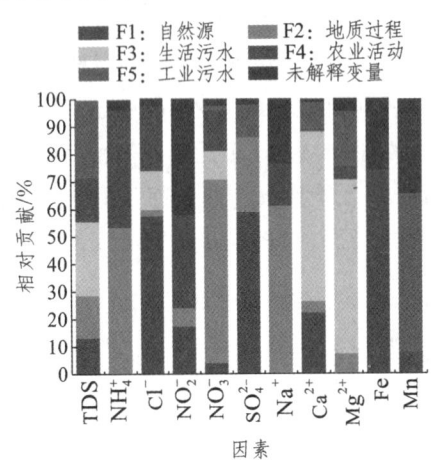

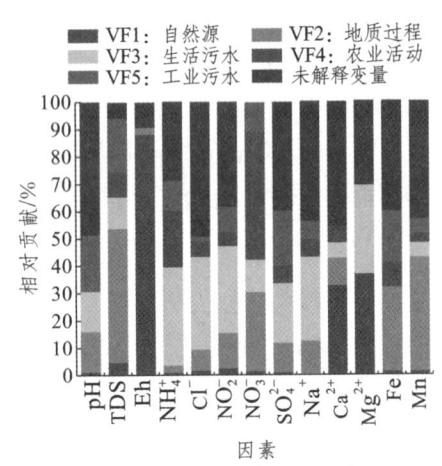

图 6-2 (a)PMF 和(b)APCS-MLR 对不同污染源理化指标的解析结果;(c)PMF 和 APCS-MLR 解析平均贡献;(d)PMF 和 APCS-MLR 解析结果拟合度比较

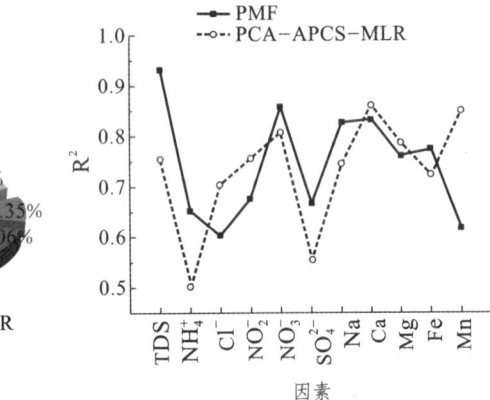

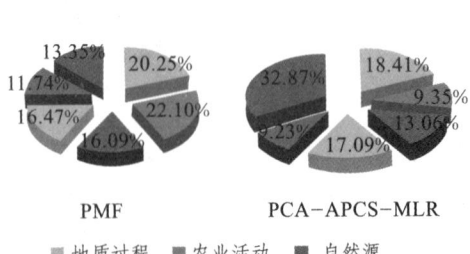

使用不同的受体模型进行源解析时可能会导致来源数量和来源类别的变化，多种受体模型配合使用，可以最大限度减少每种模型的弱点，得到更可靠的解释。如表6-1所示，CA、PCA/FA和APCS-MLR是最为常用的搭配。

单纯依靠多元统计方法来识别污染源往往会导致主观的，不确定的结果（Ren et al., 2023b）。部分学者致力于削弱多元统计模型的主观性，以获得更加准确的源解析结果。例如，Zhang et al.（2022）和Chen et al.（2022）将社会经济参数引入多元统计中，避免了污染源识别的主观性、片面性和不确定性（Xie et al., 2023; Zhang et al., 2022）；Ren et al.（2023b）利用冗余分析（Redundancy Analysis，RDA）优化了污染源识别过程，同样提高了模型的准确性。

笔者的一项研究将水化学参数（Hydrochemistry Parameters, HPs）、社会经济参数（Socioeconomic Parameters, SPs）作为多元统计中的辅助因素，结合APCS-MLR模型，用以评估污染源对流域的贡献，以求获得更准确的源解析结果（图6-3）。

如图6-3所示，采用多元统计方法，将15个HPs、12个SPs作为多元统计中的辅助因素，以实现对污染源的综合评估，定量评价潜在污染源及其对河流水污染的贡献。多元统计包括回归分析、PCA和APCS-MLR。结果表明，工业和人口增长是氨氮（NH_4^+-N）和总氮（TN）污染的主要影响因素，而总磷（TP）污染与生活污水排放和畜禽养殖的关系更为密切。基于主成分分析结果，提取了影响水化学参数（HPs）和社会经济参数（SPs）的4个潜在因素，分别解释了总方差的68.59%和82.40%。综合两组参数的主成分分析结果，污染源排序为工业废水＞农村废水＞城市污水＞浮游植物生长和农业种植。APCS-MLR的污染源分配结果显示，工业废水和农村废水的平均污染贡献率分别为35.68%和25.08%，其次是城市污水（18.73%）和浮游植物污染（15.13%），未识别污染源比例较小。社会经济参数辅助水化学参数进行多元统计，可以提高污染源识别的准确性和确定性，为决策者制定河流水质保护策略提供支持（Zhang et al., 2022）。

笔者的另一项研究考虑使用RDA分析，优化了污染源识别过程。具体地，基于RDA分析，探讨不同区域土地利用指标对水质参数的响应差异；分别利用PMF和APCS-MLR受体模型对河流污染源进行识别和分配并比较PMF和APCS-MLR模型的性能（图6-4）。

表 6-1 流域面源污染多元统计模型综合概述

国家	时间跨度	模型使用	参数选择（数量）	磷相关参数	污染源数量	参考文献
中国	2017.1—2020.7	PCA/FA; APCS-MLR	TN; TP; COD_{Cr}; COD_{Mn}; BOD… (10)	TP	4	(Xie et al., 2023)
中国	2017—2018	CA; PCA/FA; APCS-MLR	TN; TP; COD_{Cr}; COD_{Mn}; BOD_5… (15)	TP	5	(Zhang et al., 2020b)
中国	2019—2021	PCA/FA; APCS-MLR	TN; TP; COD_{Cr}; SS; DO… (8)	TP	4; 4; 3	(Zhang et al., 2018)
韩国	14 次暴雨	PCA; PMF	TP; As; Al; K^+; Na^+… (20)	TP	5	(Lee et al., 2016)
巴西	2010—2017	CA; PCA	TN; TP; $Fe^{3+/2+}$; Cl^-; NH_4^+-N… (50)	TP		(Fraga et al., 2020)
中国	2003—2015	PCA	TN; TP; NH_4^+-N; COD_{Cr}; BOD_5… (7)	TP		(Wei et al., 2019)
中国	2014—2017	PMF; APCS-MLR	TN; TP; NH_4^+-N; COD_{Cr}; BOD_5… (11)	TP	PMF:5; APCS-MLR: 4	(Ren et al., 2023b)
中国	2018—2019	CA; PCA; PMF	TN; TP; NH_4^+-N; COD_{Cr}; BOD_5… (14)	TP	5; 6	(Ren et al., 2023a)
土耳其	2019.1—2019.12	CA;PCA/FA; APCS-MLR	TN; TP; PO_4^{3-}; NO_3^--N; NH_4^+-N… (19)	TP; PO_4^{3-}	4	(Varol et al., 2022)

续表

国家	时间跨度	模型使用	参数选择（数量）	磷相关参数	污染源数量	参考文献
中国	2010—2014	PCA/FA; PMF; APCS-MLR	TN; TP; PO_4^{3-}; SS; Cl^-… (16)	TP; PO_4^{3-}	5	(Liu et al., 2020)
中国	2018—2019	CA; PCA/FA; APCS-MLR	TN; TP; NH_4^+-N; COD_{Cr}; BOD_5… (12)	TP	4	(Fu et al., 2020)
中国	2010.12—2011.6	FCA; PCA; APCS-MLR	TN; TP; Cl^-; COD_{Mn}; Pb… (17)	TP	3	(Chen et al., 2013)
中国	2018.8	PCA/FA; APCS-MLR; PMF; UNMIX	Nine organophosphate esters (OPEs)	Nine OPEs	3	(Qi et al., 2020)
中国	2014—2017	PCA; APCS-MLR	TN; TP; NH_4^+-N; COD_{Mn}; BOD… (9)	TP	3	(Chen et al., 2021)
美国	2000—2014	PCA/FA; APCS-MLR; PMF	TN; TP; TSS; Chl-a; DO… (12)	TP	4; 5	(Haji Gholizadeh et al., 2016)
中国	2015.1—2016.12	PCA; APCS-MLR	TN; TP; NH_4^+-N; COD_{Mn}; BOD… (27)	TP	4	(Zhang et al., 2022)

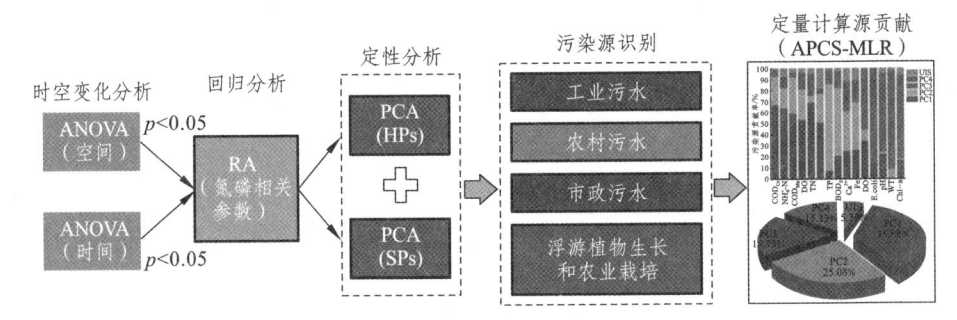

图 6-3　结合理化和社会经济参数的多元统计法识别地表水污染源（Zhang et al., 2022）

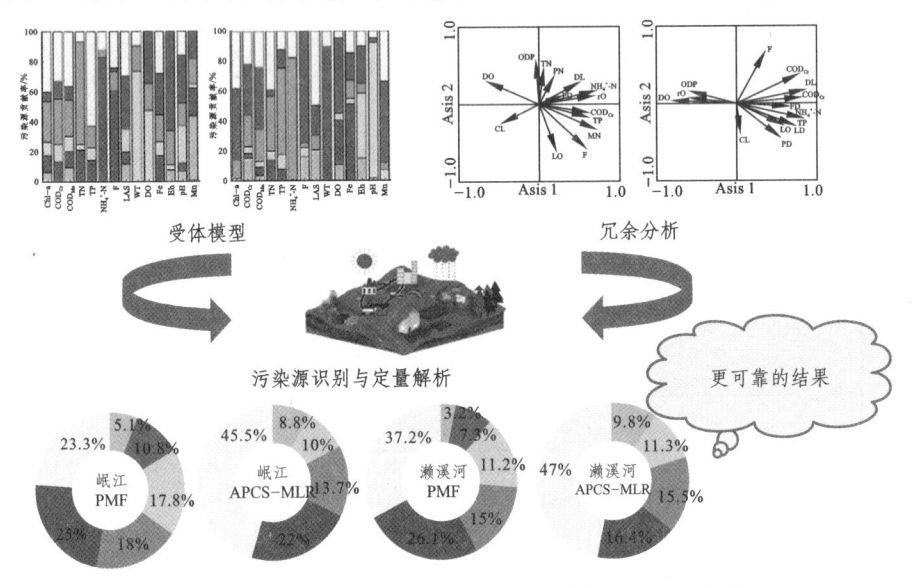

图 6-4　多元统计模型联合 RDA 分析的流域污染源解析

该研究提出了土地利用影响污染源解析的假设，选取了流量规模不同，主要污染指标不同，土地利用类型不同的两个流域（岷江、沱江的二级支流濑溪河）进行了验证。结果表明，区域间水质对土地利用的响应机制存在差异，RDA工具优化了受体模型的污染源分析过程。PMF 和 APCS-MLR 受体模型分别识别出 5 种和 4 种污染源及其相应的特征参数。在污染源分析中考虑土地利用的影响可以克服受体模型的主观性，提高污染源识别和分配的准确性。

6.2　稳定同位素法

近年来，基于稳定同位素示踪的受体模型法逐渐被用于水体污染源的定性和定量识别，通过同位素分馏作用还可揭示污染物在环境中的生物化学转化过程。该技术不受受体浓度的影响，可以通过低频率的采样实现不同污染源中污染物的分析，解析环境水体中污染物来源和贡献率，并借助稳定同位素模型的构建实现污染源识别从定性向定量研究的跨越（Zhang et al., 2020c）。

6.2.1 $\delta^{18}O_p$ 描述

通常利用稳定同位素作为示踪剂需具备以下前提：① 各潜在污染源之间存在可识别的同位素值差异；② 同位素特征值在环境中保持稳定，几乎不发生分馏作用，或者分馏作用可以识别。此外，稳定同位素分析还要求被分析元素具有两种自然存在的稳定同位素，基于样品中一种元素的单个同位素与已知比率的变化进行分析（Davies et al., 2014）。例如 $\delta^{13}C$，$\delta^{15}N$，δD 和 $\delta^{18}O_H$ 分别以 $^{13}C/^{12}C$，$^{15}N/^{14}N$，$^{2}H/^{1}H$ 和 $^{18}O/^{16}O$ 比值为基础进行稳定同位素分析（Liu et al., 2023）。磷有三种主要的同位素（如 ^{31}P，^{32}P 和 ^{33}P），其中 ^{32}P 和 ^{33}P 的半衰期分别为 14.36 和 25.34 天，只有 ^{31}P 是稳定同位素（Davies et al., 2014; Tcaci et al., 2019）。因此，磷原子不能用于稳定同位素分析。

不过，在自然环境中，磷通常只与磷酸盐分子中的 O 紧密结合。磷酸盐中的 P—O 键在地球表面环境的典型温度和压力条件下可以抵抗无机水解。P—O 键的裂解依赖于细胞内或胞外酶催化的代谢过程，这些代谢过程在生成的磷酸盐分子内赋予 O 原子一系列的平衡或动力学分馏。一方面，不同来源的磷，经过生物化学过程得到不同同位素值；另一方面，水中运输过程对 P—O 键的影响很小，其同位素信号得以保存（Tcaci et al., 2019; Wu et al., 2021）。以此为前提，基于 $^{18}O/^{16}O$ 比率的磷酸盐氧同位素（$\delta^{18}O_p$）可作为研究磷源和磷循环的良好示踪剂（Davies et al., 2014）。

6.2.2 $\delta^{18}O_p$ 分馏

$\delta^{18}O_p$ 的分馏作用将不同来源的同位素值区分开来，使其满足作为示踪剂的第一前提。同位素的分馏作用本质是一种同位素优先于另一种同位素进入反应产物（Davies et al., 2014）。$\delta^{18}O_p$ 分馏作用通常包括动力学分馏和平衡分馏。动力学分馏是指在单向反应中，由于反应速率较快，优先选择一种同位素（通常为较轻同位素）的过程；而平衡分馏是一种热力学效应，系统经过长时间的连续交换同位素，以达到最低能量系统，其中较重的同位素形成较强的键（Davies et al., 2014）。

在自然环境中，磷酸盐中的P—O键通常是稳定的。如果没有生物作用，磷酸盐和水之间的同位素交换可以忽略不计，这使得$\delta^{18}O_p$满足作为示踪剂的第二前提。$\delta^{18}O_p$在流域运输过程中发生的动力学或平衡分馏用$E\delta^{18}O_p$表示，可以通过经验方程估算（Liu et al.，2023）。

$$T = 111.4 - 4.3(E\delta^{18}O_p - \delta^{18}O_H) \tag{6-6}$$

式中，T为温度，℃；$\delta^{18}O_H$为地表水氧同位素组成，‰。

在一些非生物反应的初始阶段，P_i与固相氧化铁的吸附或磷灰石的形成可能会发生$\delta^{18}O_p$的动力学分馏，但这不会持续进行。并且随着时间推移，固相P_i中的$\delta^{18}O_p$组成将接近水相。相比之下，酶催化过程会裂解P—O键，导致P_i中的$\delta^{18}O_p$与细胞内或胞外环境中流体的$\delta^{18}O_p$之间发生动力学或平衡分馏，一些在细胞内、外环境代谢过程中可能发生的主要同位素分馏效应见图6-5（Davies et al.，2014）。

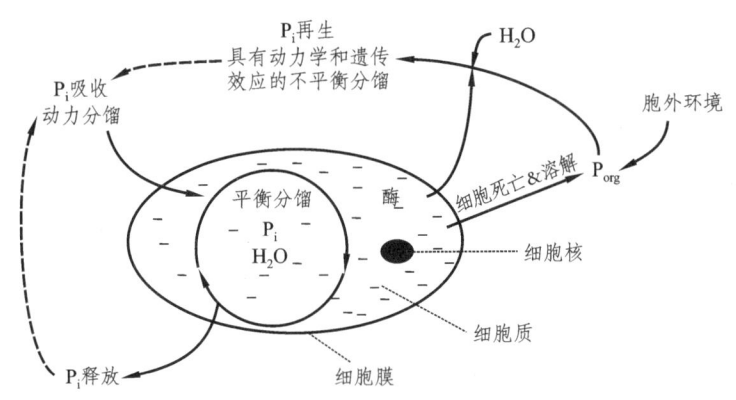

图6-5 细胞内、外环境代谢过程中可能发生的主要同位素分馏效应
（Davies et al.，2014）

如图6-5所示，水生生态系统中存在多种含磷化合物，包括P_i和P_o。其中，P_i是生物优先利用的形态，因为它能够在细胞质和细胞膜中扩散（Bjorkman and Karl，1994；Liang and Blake，2006）。因此，P_i是许多细胞内反应所必需的，包括通过磷酸化和去磷酸化（Blake et al.，1997，2005）。催化胞内P循环的主要酶是无机焦磷酸酶，它能够介导可逆反应，如ATP与ADP的相互转化（Blake et al.，2005）。由无机焦磷酸酶催化的细胞内反应导致P_i中的O原子与周围水分子中的O原子之间进行完整且非常迅速（以分钟为单位）的$\delta^{18}O_p$交换，并将O原子提供给新产生的P_i分子，由此发生动力学分馏（Blake et al.，2005）。此外，在

双重机制的驱动下，磷从生物质释放到细胞外环境。一方面，活的细胞将生物质作为副产物进行摄取和细胞内代谢反应，以维持细胞内的 P 浓度（Cooperman et al., 1992）；另一方面，在细胞死亡和裂解后。生物质中的 P_i 通过被广泛吸收和循环，由此发生平衡分馏。细胞外环境中最初的 $\delta^{18}O_p$ 组成可以反映 P 的来源，P_i 的平衡分馏释放到细胞外环境后可能会叠加。

6.2.3 $\delta^{18}O_p$ 样品预处理和测定

目前，对于 $\delta^{18}O_p$ 样品的测定，常用的方法是将水中的磷酸盐转化为高纯 Ag_3PO_4 固体，通过元素分析-同位素比质谱法（EA-IRMS），在高温热解条件下测量 $\delta^{18}O_p$ 组成（Liu et al., 2021）。在测量样品 $\delta^{18}O_p$ 之前，需要对水样中的磷进行富集，提取和纯化（Yuan et al., 2022）。目前应用最多的方法是 McLaughlin 法（图 6-6）（McLaughlin et al., 2004; Wu et al., 2021）。

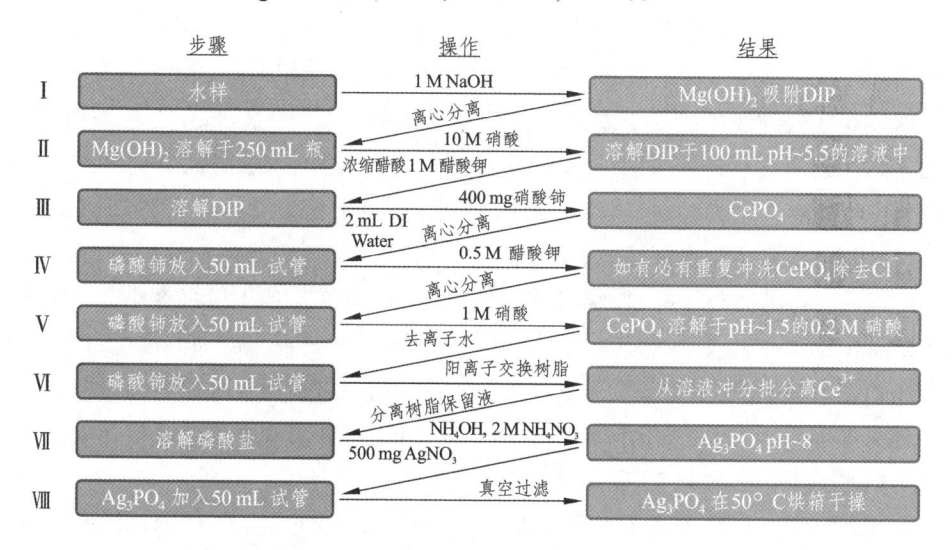

步骤		操作	结果
I	水样	1 M NaOH / 离心分离	$Mg(OH)_2$ 吸附 DIP
II	$Mg(OH)_2$ 溶解于 250 mL 瓶	10 M 硝酸 / 浓缩醋酸 1 M 醋酸钾	溶解 DIP 于 100 mL pH~5.5 的溶液中
III	溶解 DIP	400 mg 硝酸铈 / 2 mL DI Water 离心分离	$CePO_4$
IV	磷酸铈放入 50 mL 试管	0.5 M 醋酸钾 / 离心分离	如有必要重复冲洗 $CePO_4$ 除去 Cl
V	磷酸铈放入 50 mL 试管	1 M 硝酸 / 去离子水	$CePO_4$ 溶解于 pH~1.5 的 0.2 M 硝酸
VI	磷酸铈放入 50 mL 试管	阳离子交换树脂 / 分离树脂保留液	从溶液冲分批分离 Ce^{3+}
VII	溶解磷酸盐	NH_4OH, 2 M NH_4NO_3 / 500 mg $AgNO_3$	Ag_3PO_4 pH~8
VIII	Ag_3PO_4 加入 50 mL 试管	真空过滤	Ag_3PO_4 在 50°C 烘箱干燥

图 6-6　McLaughlin et al.（2004）中关于在海水中提取 DIP 并测定 $\delta^{18}O_p$ 描述的步骤

如图 6-6 所示，首先，在 8 L 磷酸盐浓度大于 0.8 μM 的海水中（较低浓度需要更大体积的海水），加入 150 mL 1 M 的氢氧化钠（NaOH）溶液。然后去掉上清液（通常在瓶子里留下 2 L），保留生成的氢氧化镁（$Mg(OH)_2$）沉淀。将此海水/$Mg(OH)_2$ 混合物转移到 250 mL 瓶中，每个样品在 3 500 r/min 下离心 10 min，再次去掉上清液。最终结果是在 250 mL 的瓶子中大约有 100 mL 的湿 $Mg(OH)_2$。

其次，将 $Mg(OH)_2$ 沉淀物溶解在 5 mL 浓乙酸中，需要 10 M 硝酸至少 5 mL。然后将溶液用 10 mL 1 M 醋酸钾缓冲至 pH 5.5。在溶液中加入 400 mg 溶解于 3 mL 水（新鲜配制）的硝酸铈（$Ce(NO_3)_3$），沉淀出白色至奶油色的磷酸铈（$CePO_4$）。样品在室温下放置 1 至 3 小时以完成沉淀。如果这个反应不完全，氧同位素可能会发生分馏。因此，$Ce(NO_3)_3$ 的添加量应足以确保铈离子（Ce^{3+}）过量存在。此外，在微酸性 pH 下进行醋酸缓冲是必要的，原因包括：① 它可以防止铈和其他干扰离子如铁和镁的水解；② 保持 $CePO_4$ 的絮凝，$CePO_4$ 的晶体很小，几乎没有 XRD 特征，无法离心，也不容易过滤，$CePO_4$ 在弱酸性溶液中的可溶性可以忽略不计，因此在弱酸性 pH 下进行缓冲可以防止在冲洗过程中沉淀的溶解。

再次，将 $CePO_4$ 逐渐转移到 50 mL 离心管中，3 500 r/min 离心 15 min 分离固体。250 mL 瓶用醋酸钾溶液冲洗 3 次，确保 $CePO_4$ 完全转移到离心管中。然后用 20 mL 0.5 M 醋酸钾反复冲洗 $CePO_4$，离心至所有氯离子（Cl^-）从上清液中去除。这需要通过在上清液中加入硝酸银并测试氯化银（AgCl）沉淀来证明。通常需要 3 次冲洗才能完全去除 Cl^-。在每次离心之间，将 $CePO_4$ 摇匀，使沉淀物再悬浮到漂洗液中。注意，$CePO_4$ 不应在去离子水中漂洗，因为没有醋酸钾，$CePO_4$ 会溶解，无法离心。随后，将保留的沉淀物溶解在 2~4 mL 的 1 M 硝酸中，并将酸的摩尔浓度降至 0.2 以下。

然后，将溶液与 4 mL 阳离子交换树脂混合，并在摇床上振荡 8 小时，以去除溶液中的 Ce^{3+}。使用玻璃柱从树脂中洗脱样品，并在第二组 50 mL 离心管中收集溶液。在溶液中加入 2 滴溴百里酚蓝指示剂，指示剂溶液颜色应为黄色。在样品中加入 1 mL 氢氧化铵（NH_4OH）和 1 mL 3 M 硝酸铵（溶液应为蓝色）。用 3 M 硝酸将 pH 调至 7（指示液应为绿色）。

最后，在 2 mL 去离子水中加入 0.5 g 硝酸银（$AgNO_3$），快速沉淀磷酸银（Ag_3PO_4）。$AgNO_3$ 的加入会降低 pH 值，指示剂溶液的颜色会变黄。明亮的黄色沉淀立即形成，并可能随着时间的推移变成黑色。在这个阶段，可以用通用 pH 试纸检查样品溶液的 pH 值，确保 pH 值约为 7（±0.5）。pH 值大于或小于 7 将导致 Ag_3PO_4 的不完全沉淀。然后将 Ag_3PO_4 真空过滤到聚碳酸酯过滤器上，并用去离子水冲洗几次。过滤器在 50 ℃ 下放置在烤箱中过夜以去除多余的水。由此获得提取的 $\delta^{18}O_p$ 样品，通过元素分析-同位素比质谱法（EA-IRMS），在高温热解条件下测量 $\delta^{18}O_p$ 组成。

不过，McLaughlin 法最早设计用于海水中无机磷酸盐的提取。在淡水流域中，该方法的应用受到了限制。原因是淡水样品中的无机或有机化合物会干扰磷酸盐的沉淀，需要通过更多的纯化步骤来克服（Ishida et al., 2019; Wu et al., 2021）。并且天然水体中磷含量较低，需要大量水样（甚至高达 1 000 L）预浓缩磷酸盐，并通过如图 6-6 所示的复杂步骤去除杂质，以求得足够量的 Ag_3PO_4 进行 $\delta^{18}O_p$ 测定，非常耗时费力（Liu et al., 2021）。例如，天然水样中各种杂质（如阴离子/阳离子、DOM、金属和微量元素）的含量高于磷酸盐，不完全去除杂质可能导致含磷化合物的水解，影响 $\delta^{18}O_p$ 测量的可靠性（Liu et al., 2021）；$CePO_4$ 沉淀物需要多次清洗（甚至高达 10 次）才能完全去除 Cl^-，繁杂的步骤容易造成磷酸盐的损失，如何保证其回收率成为新的挑战。

目前，对于 $\delta^{18}O_p$ 样品的预处理，缺乏标准化的富集、提取和纯化方法，且当前存在的方法均面临操作繁琐，潜在误差大，回收率低的难题（Liu et al., 2021; Yuan et al., 2022）。为进一步推广稳定同位素示踪法在磷污染方面的应用，还需进一步降低预处理成本，提高 $\delta^{18}O_p$ 测定的准确性。Liu et al.（2021）认为常用的方法通常为瞬时采样，对于获得的样品存在一定的偶然性，其开发出一种原位富集、洗脱和纯化磷酸盐的方法。将 Zr-Oxide 凝胶置于水体中 15 天，用于原位吸附磷酸盐（Liu et al., 2021）。该方法得到了较为准确的 $\delta^{18}O_p$ 值，不过耗时较长的问题可能会限制其应用；Tcaci et al.（2019）针对低浓度淡水样品，开发出一种扭转纺纱模式，能够在大约 24 小时内，现场预处理超过 1 000 L P 浓度＜0.016 mg/L 的样品。结合一种新的冷冻干燥方法，以最大限度地提高产率，最大限度地减少磷酸银污染。

6.2.4 $\delta^{18}O_p$ 污染源解析

由于不同污染源 $\delta^{18}O_p$ 同位素存在差异，且 $\delta^{18}O_p$ 在水体中相对稳定，因此可以用于 P 污染源解析（Dang et al., 2018; Gross et al., 2016; Ishida et al., 2019; Li et al., 2021; Mingus et al., 2019; Yuan et al., 2022）。$\delta^{18}O_p$ 可以区分不同的潜在磷源，但单独使用尚不能定量估计每个污染源的贡献率，还需引入同位素混合模型进行定量估算。主要的模型包括 IsoError、IsoSource、SIAR 和 MixSIAR 模型（Liu et al., 2023）。例如 Wu et al.（2021）使用 MixSIAR 模型解析了污染源对降雨中 P 的贡献率，结果表明，在旱季煤燃烧、生物质燃烧和土壤对 P 的贡献率分别为 39%±16%、22%±18%和 17%±13%，而在雨季海洋源和煤燃烧分别贡献 22%±19% 和 24%±12%（Wu et al., 2021）。

基于近年来关于 $\delta^{18}O_p$ 稳定同位素的文献（Gooddy et al., 2016; Granger et al., 2017; Huang et al., 2022; Ishida et al., 2019; Liu et al., 2023; Pistocchi et al., 2017; Tonderski et al., 2017; Yang et al., 2018; Yuan et al., 2019; Yuan et al., 2022），总结了不同流域和不同污染源 $\delta^{18}O_p$ 值，如图 6-7（a）和（b）所示（彩图见附录 2）。

（a）

（b）

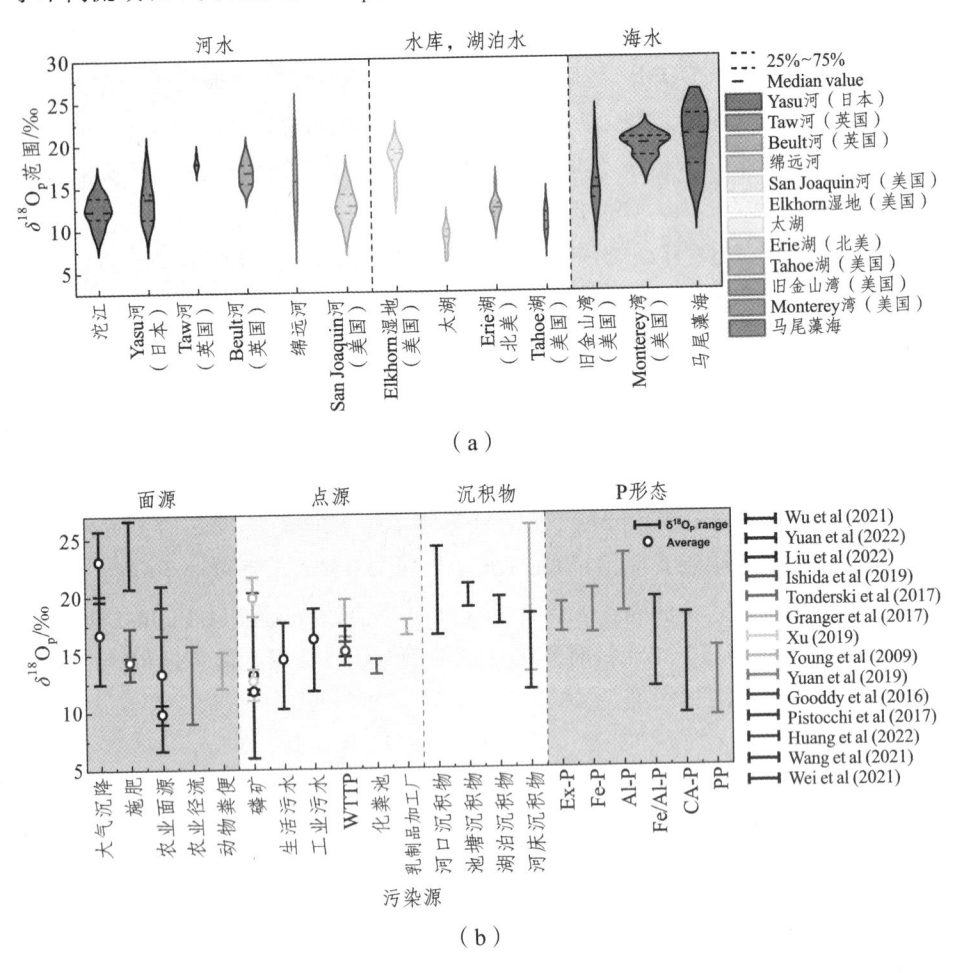

图 6-7 （a）不同流域 $\delta^{18}O_p$ 值范围小提琴图；（b）不同污染源 $\delta^{18}O_p$ 值区间

如图 6-7（a）所示，对于同一流域的不同污染源，$\delta^{18}O_p$ 宽且重叠范围较大，仅用 $\delta^{18}O_p$ 识别磷源可能较困难。通常情况下，对于同位素混合模型，使用两种或两种以上的示踪剂可以获得较为准确的结果。例如，Zhang et al.（2020c）在

SIAR 模型中使用了 $\delta^{15}N_{NO_3^-}$ 和 $\delta^{18}O_{NO_3^-}$ 准确解析了成都平原多土地利用区硝酸盐来源（图 6-8）。

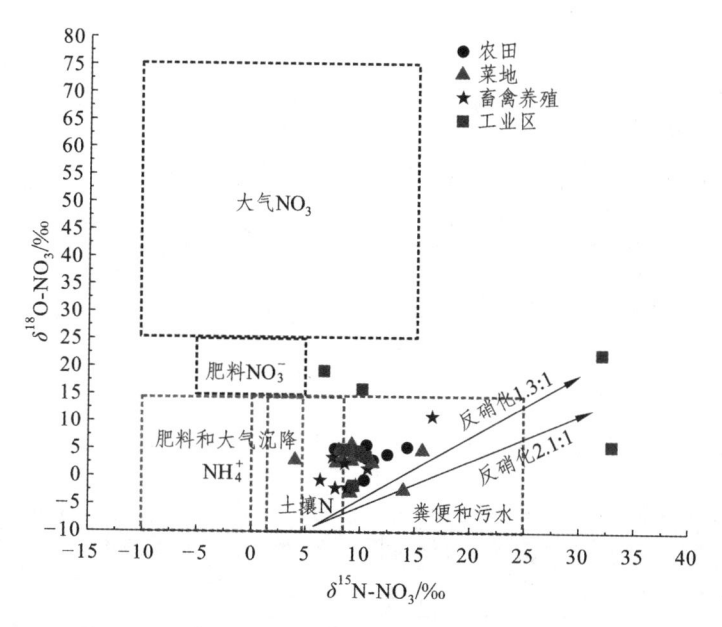

图 6-8　$\delta^{15}N_{NO_3^-}$ 和 $\delta^{18}O_{NO_3^-}$ 准确解析了成都平原多土地利用区硝酸盐来源
（Zhang et al., 2020c）

然而，这种方法对于 $\delta^{18}O_P$ 并不适用，因为 P 仅有一种稳定同位素，无法作为示踪剂。因此，有研究人员提出，引入其他可靠的稳定同位素工具，与 $\delta^{18}O_P$ 耦合进行源解析（Li et al., 2021）。例如，Liu et al.（2023）引入了 $\delta^{13}C$、$\delta^{15}N$ 和 δD，分别使用 SIAR 和 IsoSource 模型估算了沱江潜在磷源的比例贡献。$\delta^{13}C$、$\delta^{15}N$ 和 δD 等稳定同位素的引入使得 $\delta^{18}O_P$ 在散点图中区分开来（图 6-9）。不过，还需进一步验证这几种同位素与 $\delta^{18}O_P$ 来源的相关性。

如图 6-7（a）所示，对于不同流域，$\delta^{18}O_P$ 整体范围存在一定的差异。如图 6-7（b），不同流域同一种污染物的 $\delta^{18}O_P$ 也存在较大差异。尽管 $\delta^{18}O_P$ 能将同一流域的不同污染物区分开来，但是其应用存在区域局限性。因此，建立区域性的 $\delta^{18}O_P$ 数据库，对 $\delta^{18}O_P$ 作为示踪剂的稳定同位素溯源法的推广具有积极意义。

综上所述，稳定同位素溯源法在磷污染源解析方面具有直接、高效的优势。然而，污染源同位素特征值重叠，同位素分馏效应以及磷污染输入时空差异等内、外部因素都会带来的识别误差和不确定性，都将影响和限制稳定同位素溯源法的准确性和实际应用。随着淡水稳定同位素 $\delta^{18}O_P$ 检测方法的改进和简化，以及相关模型算法的开发，磷酸盐氧同位素示踪法在识别水体磷污染和揭示磷素迁移转化规律中将具更大潜力。

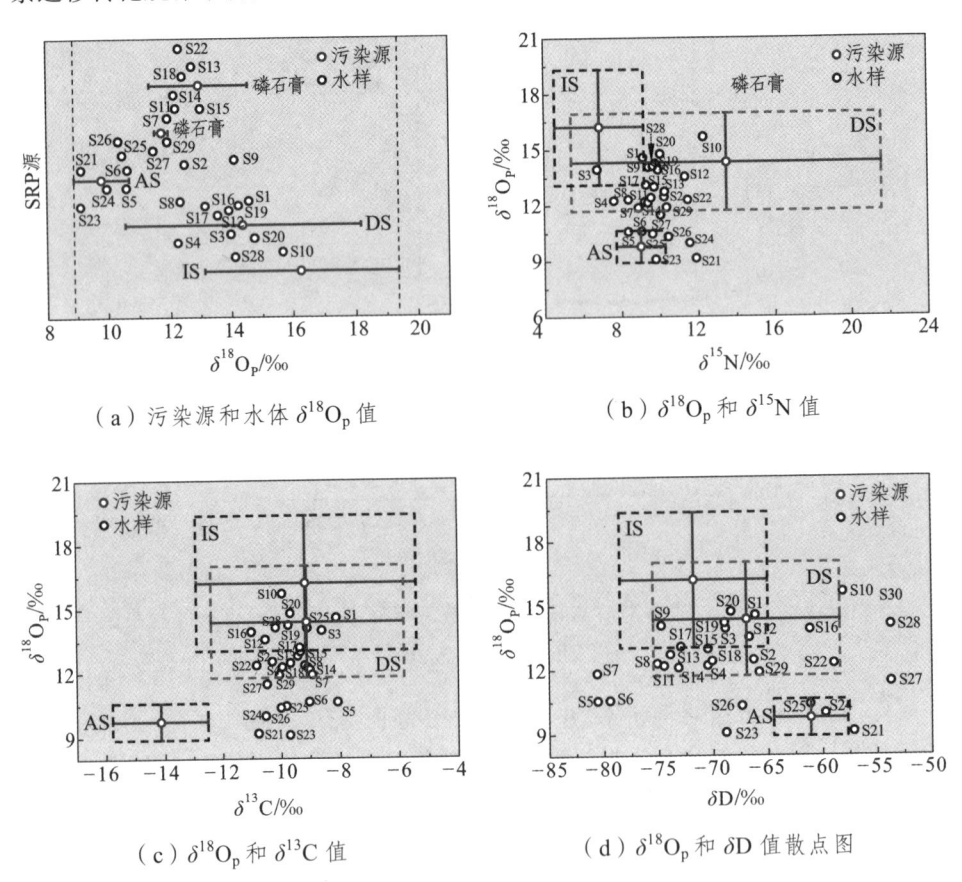

图 6-9 沱江（a）污染源和水体 $\delta^{18}O_P$ 值；（b）$\delta^{18}O_P$ 和 $\delta^{15}N$ 值；（c）$\delta^{18}O_P$ 和 $\delta^{13}C$ 值；（d）$\delta^{18}O_P$ 和 δD 值散点图（IS 为工业污水；DS 为生活污水；AS 为农业污水）
（Liu et al., 2023）

参考文献

[1] Bjorkman K, Karl D M. 1994. Bioavailability of inorganic and organic phosphorus-compounds to natural assemblages of microorganisms in Hawaiian coastal waters[J]. Mar. Ecol. Prog. Ser. 111(3), 265-273.

[2] Blake R E, O'Neil J R, Surkov A V. 2005. Biogeochemical cycling of phosphorus: Insights from oxygen isotope effects of phosphoenzymes[J]. Am. J. Sci. 305(6-8), 596-620.

[3] Chen H, Teng Y, Yue W, et al. 2013. Characterization and source apportionment of water pollution in Jinjiang River, China[J]. Environ. Monit. Assess. 185(11), 9639-9650.

[4] Chen K, Liu Q, Jiang Q, et al. 2022. Source apportionment of surface water pollution in North Anhui Plain, Eastern China, using APCS-MLR model combined with GIS approach and socioeconomic parameters[J]. Ecol. Indic. 143, 109324.

[5] Chen S, Wang S, Yu Y, et al. 2021. Temporal trends and source apportionment of water pollution in Honghu Lake, China[J]. Environ. Sci. Pollut. Res. Int. 28(42), 60130-60144.

[6] Cooperman B S, Baykov A A, Lahti R. 1992. Evolutionary conservation of the active-site of soluble inorganic pyrophosphatase[J]. Trends Biochem. Sci. 17(7), 262-266.

[7] Dang D H, Evans R D, Durrieu G, et al. 2018. Quantitative model of carbon and nitrogen isotope composition to highlight phosphorus cycling and sources in coastal sediments (Toulon Bay, France)[J]. Chemosphere 195, 683-692.

[8] Davies C L, Surridge B W J, Gooddy D C. 2014. Phosphate oxygen isotopes within aquatic ecosystems: global data synthesis and future research priorities[J]. Sci. Total Environ. 496, 563-575.

[9] Fraga M S, Reis G B, da Silva D D, et al. 2020. Use of multivariate statistical methods to analyze the monitoring of surface water quality in the Doce River basin, Minas Gerais, Brazil[J]. Environ. Sci. Pollut. Res. Int. 27(28), 35303-35318.

[10] Fu D, Wu X, Chen Y, et al. 2020. Spatial variation and source apportionment of surface water pollution in the Tuo River, China, using multivariate statistical techniques[J]. Environ. Monit. Assess. 192(12), 745.

[11] Gooddy D C, Lapworth D J, Bennett S A, et al. 2016. A multi-stable isotope framework to understand eutrophication in aquatic ecosystems[J]. Water Res. 88, 623-633.

[12] Granger S J, Heaton T H E, Pfahler V, et al. 2017. The oxygen isotopic composition of phosphate in river water and its potential sources in the Upper River Taw catchment, UK[J]. Sci. Total Environ. 574, 680-690.

[13] Gross A, Turner B L, Goren T, et al. 2016. Tracing the sources of atmospheric phosphorus deposition to a tropical rain forest in Panama using stable Oxygen Isotopes[J]. Environ. Sci. Technol. 50(3), 1147-1156.

[14] Haji Gholizadeh M, Melesse A M, Reddi L. 2016. Water quality assessment and apportionment of pollution sources using APCS-MLR and PMF receptor modeling techniques in three major rivers of South Florida[J]. Sci. Total Environ. 566-567, 1552-1567.

[15] Huang T, Wang J, Xu Z, et al. 2022. Phosphate oxygen isotope in river sediments and its potential sources in Chaohu watershed, China[J]. J. Soils Sediments 22(5), 1585-1596.

[16] Ishida T, Uehara Y, Iwata T, et al. 2019. Identification of phosphorus sources in a watershed using a phosphate oxygen isoscape approach[J]. Environ. Sci. Technol. 53(9), 4707-4716.

[17] Lee D H, Kim J H, Mendoza J A, et al. 2016. Characterization and source identification of pollutants in runoff from a mixed land use watershed using ordination analyses[J]. Environ. Sci. Pollut. Res. Int. 23(10), 9774-9790.

[18] Li Q, Yuan H, Li H, et al. 2021. Tracing the sources of phosphorus along the salinity gradient in a coastal estuary using multi-isotope proxies[J]. Sci. Total Environ. 792, 148353.

[19] Liang Y, Blake R E. 2006. Oxygen isotope signature of P_i regeneration from organic compounds by phosphomonoesterases and photooxidation[J]. Geochim. Cosmochim. Acta 70(15), 3957-3969.

[20] Liu D, Li X, Zhang Y, et al. 2023. Using a multi-isotope approach and isotope mixing models to trace and quantify phosphorus sources in the Tuojiang River, Southwest China[J]. Environ. Sci. Technol. 57(19), 7328-7335.

[21] Liu L, Dong Y, Kong M, et al. 2020. Insights into the long-term pollution trends and sources contributions in Lake Taihu, China using multi-statistic analyses models[J]. Chemosphere 242, 125272.

[22] Liu Y, Wang J, Chen J, et al. 2021. Method for phosphate oxygen isotopes analysis in water based on in situ enrichment, elution, and purification[J]. J. Environ. Manage. 279, 111618.

[23] McLaughlin K, Silva S, Kendall C, et al. 2004. A precise method for the analysis of $\delta^{18}O$ of dissolved inorganic phosphate in seawater[J]. Limnol. Oceanogr. Meth. 2(7), 202-212.

[24] Mingus K A, Liang X M, Massoudieh A, et al. 2019. Stable isotopes and bayesian modeling methods of tracking sources and differentiating bioavailable and recalcitrant phosphorus pools in suspended particulate matter[J]. Environ. Sci. Technol. 53(1), 69-76.

[25] Pistocchi C, Tamburini F, Gruau G, et al. 2017. Tracing the sources and cycling of phosphorus in river sediments using oxygen isotopes: Methodological adaptations and first results from a case study in France[J]. Water Res. 111, 346-356.

[26] Qi Y, Liu X, Wang Z, Yao Z, et al. 2020. Comparison of receptor models for source identification of organophosphate esters in major inflow rivers to the Bohai Sea, China[J]. Environ. Pollut. 265(Pt B), 114970.

[27] Ren X, Yang C, Zhao B, et al. 2023a. Water quality assessment and pollution source apportionment using multivariate statistical and PMF receptor modeling techniques in a sub-watershed of the upper Yangtze River, Southwest China[J]. Environ. Geochem. Health 45(9), 6869-6887.

[28] Ren X, Zhang H, Xie G, et al. 2023b. New insights into pollution source analysis using receptor models in the upper Yangtze river basin: Effects of land use on source identification and apportionment[J]. Chemosphere 334, 138967.

[29] Tcaci M, Barbecot F, Helie J F, et al. 2019. A new technique to determine the phosphate oxygen isotope composition of freshwater samples at low ambient phosphate concentration[J]. Environ. Sci. Technol. 53(17), 10288-10294.

[30] Tonderski K, Andersson L, Lindstrom G, et al. 2017. Assessing the use of delta(18)O in phosphate as a tracer for catchment phosphorus sources[J]. Sci. Total Environ. 607-608, 1-10.

[31] Varol M, Karakaya G, Alpaslan K. 2022. Water quality assessment of the Karasu River (Turkey) using various indices, multivariate statistics and

APCS-MLR model[J]. Chemosphere 308(Pt 2), 136415.

[32] Wei H W, Hassan M, Che Y, et al. 2019. Spatio-Temporal Characteristics and Source Apportionment of Water Pollutants in Upper Reaches of Maotiao River, Southwest of China, from 2003 to 2015[J]. J. Environ. Inform.

[33] Wu Y, Huang X, Jiang Z, et al. 2021. Identifying sources of phosphorus in precipitation using phosphate oxygen isotope in a human and monsoon Co-affected embayment[J]. Atmospheric Environ[J]. 244, 118008.

[34] Xie D, Li X, Zhou T, et al. 2023. Estimating the contribution of environmental variables to water quality in the postrestoration littoral zones of Taihu Lake using the APCS-MLR model[J]. Sci. Total Environ. 857(Pt 3), 159678.

[35] Yang Z, Cheng B, Xu Y, et al. 2018. Stable isotopes in water indicate sources of nutrients that drive algal blooms in the tributary bay of a subtropical reservoir[J]. Sci. Total Environ. 634, 205-213.

[36] Yuan H, Li Q, Kukkadapu R K, et al. 2019. Identifying sources and cycling of phosphorus in the sediment of a shallow freshwater lake in China using phosphate oxygen isotopes[J]. Sci. Total Environ. 676, 823-833.

[37] Yuan H, Wang H, Dong A, et al. 2022. Tracing the sources of phosphorus in lake at watershed scale using phosphate oxygen isotope $\delta^{18}O_P$. Chemosphere 305, 135382.

[38] Zhang H, Cheng S Q, Li H F, et al. 2020a. Groundwater pollution source identification and apportionment using PMF and PCA-APCA-MLR receptor models in a typical mixed land-use area in Southwestern China[J]. Sci. Total Environ. 741, 140383.

[39] Zhang H, Li H, Gao D, et al. 2022. Source identification of surface water pollution using multivariate statistics combined with physicochemical and socioeconomic parameters[J]. Sci. Total Environ. 806(Pt 3), 151274.

[40] Zhang H, Li H, Yu H, et al. 2020b. Water quality assessment and pollution source apportionment using multi-statistic and APCS-MLR modeling techniques in Min River Basin, China[J]. Environ. Sci. Pollut. Res. Int. 27(33), 41987-42000.

[41] Zhang H, Xu Y, Cheng S Q, et al. 2020c. Application of the dual-isotope approach and Bayesian isotope mixing model to identify nitrate in groundwater

of a multiple land-use area in Chengdu Plain, China[J]. Sci. Total Environ. 717, 137134.

[42] Zhang L, Zhu G, Ge X, et al. 2018. Novel insights into heavy metal pollution of farmland based on reactive heavy metals (RHMs): Pollution characteristics, predictive models, and quantitative source apportionment[J]. J. Hazard. Mater. 360, 32-42.

第 7 章

磷的源解析方法综合评述

前面三章依次对清单分析法、扩散模型法和受体模型法进行了介绍，并对这些模型的研究现状进行了回顾性分析。本章将对上述模型进行对比总结，包括对模型特性进行定量分析，并基于前文的讨论对各模型进行打分。

7.1 模型特性定量分析

基于前面章节对磷的源解析方法研究现状的回顾性分析，将相关文献中的研究区域范围、研究时间跨度和模型应用场景等数据提取出来进行了量化分析，如图7-1所示（彩图见附录2）。

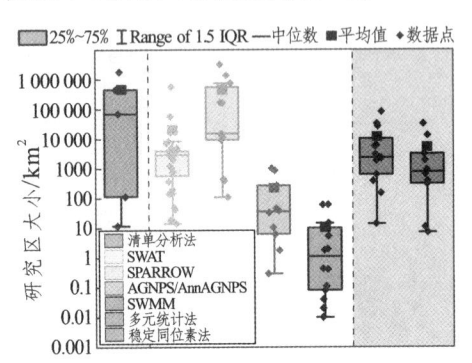

（a）不同溯源方法相关研究中研究区大小统计

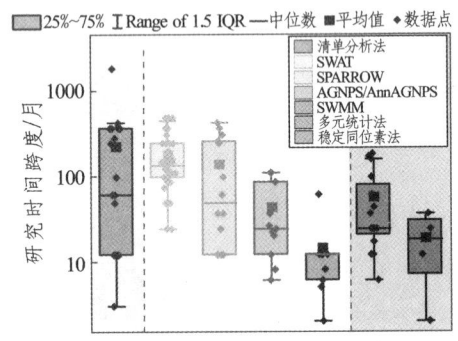

（b）不同溯源方法相关研究中时间跨度统计

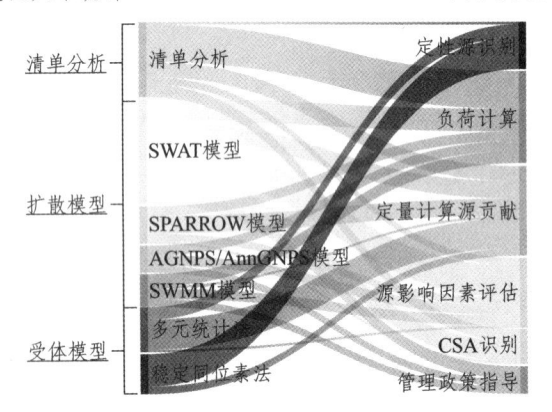

（c）不同溯源方法对应应用场景

图7-1 （a）不同溯源方法相关研究中研究区大小统计；（b）不同溯源方法相关研究中时间跨度统计；（c）不同溯源方法对应应用场景

图7-1（a）和（b），对于清单分析法，其研究区域范围和研究时间跨度都是

所有溯源方法中最大的；在扩散模型中，SWAT 综合性最强，其研究区域范围和研究时间跨度均处于四种模型的中等水平。SPARROW 侧重于体现磷源的空间异质性，在时间分辨率上较弱，其流域范围通常较大。AGNPS/AnnAGNPS 和 SWMM 均是基于事件的流域模型，其研究区域范围和研究时间跨度均相对较小；对于受体模型，多元统计侧重于表现流域整体污染情况，时间跨度相对较大。稳定同位素多为瞬时采样，其时间跨度往往较小。

当前主要的磷溯源模型最终的应用场景如图 7-1（c）所示。将应用程度由浅入深分为定性源识别、负荷计算、定量计算源贡献、源影响因素评估、CSA 识别和指导管理措施六个方面。可以看到清单分析法、SPARROW、AGNPS/AnnAGNPS、SWMM 和多元统计等方法应用场景仅为负荷计算和定量源解析。而扩散模型中的 SWAT 模型应用场景最广泛，深入到了源影响因素评估、CSA 识别和指导管理措施等方面。目前对于稳定同位素法，由于淡水中 $\delta^{18}O_p$ 的难以富集和另一个稳定示踪剂缺失这两个问题尚未得到妥善解决，还主要停留在磷源的定性识别上（Liu et al., 2021; Yuan et al., 2022）。

7.2　源解析方法评分

这些模型具有各自的特点，不存在适用于所有应用场景的"最佳模型"。这是由于模型在复杂性、数据要求以及最终提供的输出等方面存在差异。选择模型时应考虑研究最终目标和需要产出的尺度，需仔细审查所需数据的广度和质量以及流域的物理特征（Abdelwahab et al., 2018）。为此，基于前文对各溯源模型的讨论，从模型复杂性、数据要求、空间分辨率、输出尺度、时间分辨率、时间跨度和应用程度等七个方面对溯源模型进行了评价，见表 7-1。需要说明的是，以下评分是基于前面章节对这些方法的讨论和 7.1 节中定量分析得出，存在一定的主观性，仅供参考。

7.3　关于源解析方法与磷形态的讨论

最后，统计了不同溯源方法对应的磷形态。如图 7-2 所示（彩图见附录 2），绝大部分源解析方法在使用过程中都主要针对 TP，仅有少部分研究关注到了其他磷形态。然而，在前面章节的讨论中，得出了磷对流域富营养化的影响高度依赖于磷形态，而不是 TP_w 的绝对浓度。事实上流域水体及沉积物中通常存在种类丰富的磷，不同形态的磷会对流域产生不同的影响。当前源解析方法对磷形态的忽视可能会使获得的结果难以准确评估 P 源对富营养化的具体影响。

表 7-1 流域磷面源污染方法模型综合评价

磷的源解析方法		复杂性	数据要求	空间分辨率	输出尺度	时间分辨率	时间跨度	应用程度
清单分析		★★★★★	★★★★★	★★★	★★★★★	★	★★★★★	★★★★
扩散模型	SWAT	★★★★	★★★★	★★★★	★★★	★★★	★★★	★★★★★
	SPARROW	★★★	★★★★★	★★★	★★★★	★	★★★	★★★
	AGNPS/AnnAGNPS	★★★★	★★★★★	★★★★★	★★	★★★★★	★★	★★★
	SWMM	★★★	★★★★	★★★★★	★★	★★★★★	★	★★★★
受体模型	多元统计法	★★	★★	★★★	★★★	★★★	★★★	★★★
	稳定同位素法	★★	★★★★	★★★	★★★	★★	★	★

注：根据以上关于源解析方法的讨论，定义了五个等级，其中★★★★★为最大值，★为最小值，分数只代表程度高低，不代表好坏。

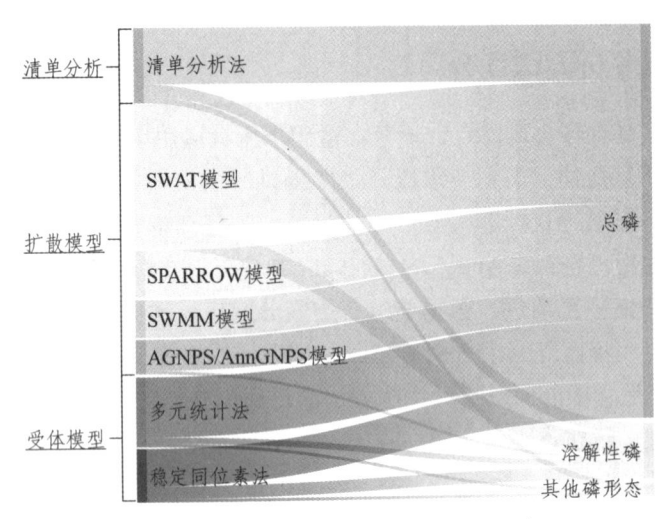

图 7-2 不同溯源方法所对应的磷形态

不同的磷形态经点源或面源进入流域，在经过分散溶解、平流运输、沉积和再悬浮等等过程后，被沉积物和水生植物捕获，经由一定的生物地球化学过程将再次进入水体。在磷污染从源到汇的迁移转化过程中，将牵涉出两个问题。首先，量化点源和面源污染中的磷形态组成；其次，量化磷在水体、水-悬浮颗粒物和水-沉积物界面的转化与释放。解决上述两个问题，并纳入源解析模型，可以得到更加准确的结果。

不过，并非所有模型都适合将磷形态纳入计算。清单分析难以获取特定形态磷在较大时空尺度下的具体排放数据；SPARROW 模型在空间尺度较大的同时对监测站数量需求也较大，若想要获取能用于 SPARROW 模型的磷形态信息，可能增加该模型的溯源成本；尽管 AGNPS/AnnAGNPS 和 SWMM 研究空间尺度较小，却需要大量集水区和地表累积数据，很难定义特定磷形态；对于多元统计法，在相对较小的城市景观水体尺度下，Zhu et al.（2023）利用 APCS-MLR 模型，量化了沉积物中 Ex-P、Fe-P、Al-P 和 Ca-P 等磷形态的贡献。该方法在较大尺度下可能难以实现对磷形态的评估，多元统计模型多用于污染源的整体评价，磷形态浓度通常是其中的一个参数。由此可见，限制溯源模型对磷形态源解析的主要因素是空间尺度和数据获取难度。

基于上述讨论以及图 7-2 对磷溯源方法和磷形态对应关系的总结，重点关注 SWAT 和稳定同位素模型在磷形态溯源的前景。SWAT 模型是目前应用最成熟的扩散模型之一。其时空跨度低于 SPARROW 模型，对子集水区和地表累积数据的要求不及 AGNPS/AnnAGNPS 和 SWMM 模型。因此，在扩散模型中，SWAT 是最有潜力对流域磷形态来源进行定量分析的模型。目前，也有研究提到使用 SWAT 量化污染源对 P_i、P_o 和 DP 等磷形态的贡献（Miralha et al., 2021; Ren et al., 2022）。在 SWAT 中，由于磷形态信息主要来自官方的监测数据，所以分类较为粗略（Millier and Hooda, 2011; Miralha et al., 2021）。还需进一步对更细致的磷形态进行分析，以量化这些磷形态对流域的贡献。

磷从源到汇的过程中，并非一成不变。经由不同的生物地球化学过程，不同的磷形态会在水体、水-悬浮颗粒物和水-沉积物界面发生复杂的迁移转化。然而，在讨论的溯源模型中，很少有研究将不同磷形态的迁移转化过程（如 PP 与 DP 的相互转化，P_o 的矿化以及沉积物中 Ex-P、Fe-P、Ca-P 和 P_o 的释放等）系统地考虑到模型计算中。实际上，SWAT 通过参数的设置可以做到描述污染物在流域中的迁移与扩散（Ni et al., 2022）。不过，转化参数的设置需要以磷的转化机理为基础（El-Khoury et al., 2015）。由于磷的转化过程较为复杂，需要进一步明确流域中的磷在水体，悬浮颗粒物和沉积物等介质中的转化机制，并量化转化过程。若能在 SWAT 中实现磷形态迁移转化的量化，势必将更准确地解析磷污染来源，为流域磷管理提供更精准的信息。

稳定同位素法具有低频采样的优势。一方面，该方法容易将 $\delta^{18}O_p$ 与特定磷形态联系起来。一些研究确定了 $\delta^{18}O_p$ 与 Ex-P、Fe/Al-P、Ca-P、P_o 和 Res-P 等磷形态的 $\delta^{18}O_p$ 的值，并建立了 $\delta^{18}O_p$ 与磷形态的相关性（Wang et al., 2021; Yuan et al., 2019; Yuan et al., 2022）。另外，Mingus et al.（2019）利用 $\delta^{18}O_p$ 联合同位素混合模型区分了 SP 中生物有效磷和难降解磷，并定性识别了其来源。由此可知，利用稳定同位素法可避免对特定磷形态的大量数据需求。

另一方面，第 6 章中提到，在使用稳定同位素进行源解析时，存在仅用 $\delta^{18}O_p$ 难以得到准确结果的问题。当前的研究通常是引入 $\delta^{13}C$，$\delta^{15}N$ 和 δD 等同位素辅佐进行示踪。同理，作为与 $\delta^{18}O_p$ 具有同源性的磷组分和形态等特征，同样可作为辅助信息用于稳定同位素源解析。Li et al.（2021）利用不同结合磷浓度辅助河口沉积物中的磷来源，不过只做到了定性识别。若与 $\delta^{13}C$，$\delta^{15}N$ 和 δD 等同位素辅佐示踪类似，将磷组分信息引入稳定同位素混合模型进行源解析，则有望提高对 TP 源解析的准确性并实现磷形态的定量源解析。

参考文献

[1] Abdelwahab O M M, Ricci G F, De Girolamo A M, et al. 2018. Modelling soil erosion in a Mediterranean watershed: Comparison between SWAT and AnnAGNPS models[J]. Environ. Res. 166, 363-376.

[2] El-Khoury A, Seidou O, Lapen D R, et al. 2015. Combined impacts of future climate and land use changes on discharge, nitrogen and phosphorus loads for a Canadian river basin[J]. J. Environ. Manage. 151, 76-86.

[3] Li Q, Yuan H, Li H, et al. 2021. Tracing the sources of phosphorus along the salinity gradient in a coastal estuary using multi-isotope proxies[J]. Sci. Total Environ. 792, 148353.

[4] Liu Y, Wang J, Chen J, et al. 2021. Method for phosphate oxygen isotopes analysis in water based on in situ enrichment, elution, and purification[J]. J. Environ. Manage. 279, 111618.

[5] Millier H K, Hooda P S. 2011. Phosphorus species and fractionation---why sewage derived phosphorus is a problem[J]. J. Environ. Manage. 92(4), 1210-1214.

[6] Mingus K A, Liang X M, Massoudieh A, et al. 2019. Stable isotopes and bayesian modeling methods of tracking sources and differentiating bioavailable

and recalcitrant phosphorus pools in suspended particulate matter[J]. Environ. Sci. Technol. 53(1), 69-76.

[7] Miralha L, Muenich R L, Scavia D, et al. 2021. Bias correction of climate model outputs influences watershed model nutrient load predictions[J]. Sci. Total Environ. 759, 143039.

[8] Ni Z, Li Y, Wang S. 2022. Cognizing and characterizing the organic phosphorus in lake sediments: Advances and challenges[J]. Water Res. 220, 118663.

[9] Ren D, Engel B, Mercado J A V, et al. 2022. Modeling and assessing water and nutrient balances in a tile-drained agricultural watershed in the U. S[J]. Corn Belt. Water Res. 210, 117976.

[10] Wang J, Huang T, Wu Q, et al. 2021. Sources and cycling of phosphorus in the sediment of rivers along a eutrophic lake in China indicated by phosphate oxygen isotopes[J]. ACS Earth Space Chem. 5(1), 88-94.

[11] Yuan H, Li Q, Kukkadapu R K, et al. 2019. Identifying sources and cycling of phosphorus in the sediment of a shallow freshwater lake in China using phosphate oxygen isotopes[J]. Sci. Total Environ. 676, 823-833.

[12] Yuan H, Wang H, Dong A, et al. 2022. Tracing the sources of phosphorus in lake at watershed scale using phosphate oxygen isotope $\delta^{18}O_P$. Chemosphere 305, 135382.

[13] Zhu Z, Wang Z, Yu Y, et al. 2023. Occurrence forms and environmental characteristics of phosphorus in water column and sediment of urban waterbodies replenished by reclaimed water[J]. Sci. Total Environ. 888, 164069.

附 录

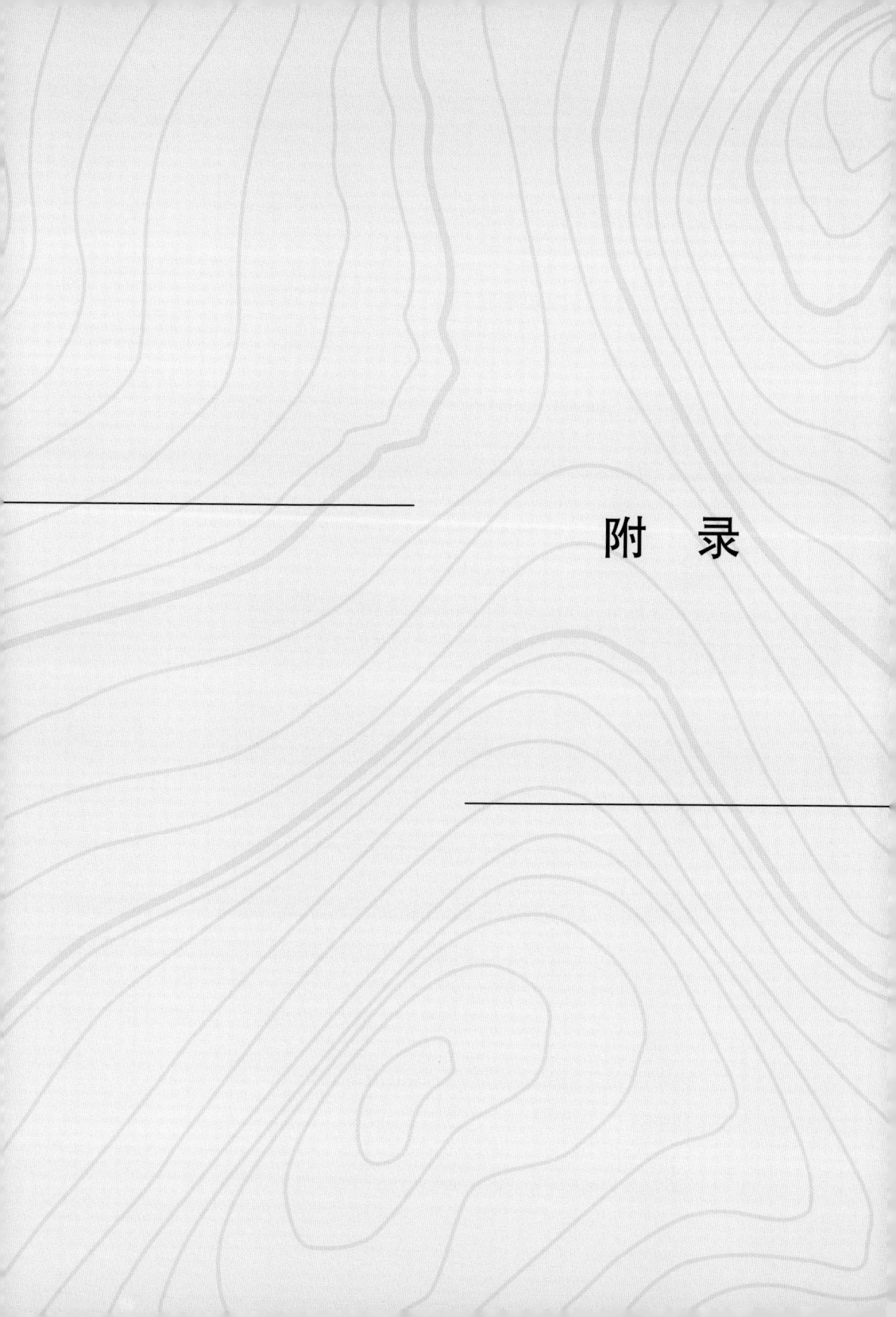

附录一 术语表（按首字母排序）

1. ^{31}P NMR（^{31}P nuclear Magnetic Resonance Spectroscopy，^{31}P 核磁共振光谱）；
2. $\delta^{18}O_p$（磷酸盐氧同位素）；
3. AGNPS model（Agricultural Non-Point Source Pollution Model，AGNPS 模型）；
4. Al-P（Al Bound Phosphorus，铝结合磷）；
5. AnnAGNPS model（Annualized Agricultural Non-Point Source Pollutant Model，AnnAGNPS 模型）；
6. APA（Alkaline Phosphatase，碱性磷酸酶）；
7. APCS-MLR（Adaptive Potential-Concentration Surface- Multiple Linear Regression，绝对主成分多元线性回归模型）；
8. Bound Phosphorus（结合态磷）；
9. BMP（Best Management Practice，最佳管理措施）；
10. Ca-P（Ca Bound Phosphorus，钙结合磷）；
11. CA（Cluster Analysis，聚类分析）；
12. Chl-a（Chlorophyll a，叶绿素 a）；
13. CN（Curve Number，曲线数）；
14. COD（Chemical Oxygen Demand，化学需氧量）；
15. DDVP（Dichlorvos, O,O-Dimethyl-O-2,2-Dichlorovinylphosphate, $C_4H_7Cl_2O_4P$，敌敌畏）；
16. DEM（Digital Elevation Model，数字高程模型）；
17. DHSVM model（Distributed Hydrology Soil Vegetation Model，DHSVM 模型）；
18. DIP（Dissolved Inorganic Phosphorus，溶解性无机磷）；
19. DOMs（Dissolved Organic Matters，溶解性有机物）；
20. DOP（Dissolved Organic Phosphorus，溶解性有机磷）；
21. DP（Dissolved Phosphorus，溶解态磷）；
22. Diester-P（二酯磷）；
23. E_{ns}（Nash-Sutcliffe efficiency Coefficient，Nash-Sutcliffe 效率系数）；
24. EDTA（Ethylene diamine tetraacetate，乙二胺四乙酸酯）；

25. EEM（Excitation—Emission—Matrix Spectra，三维荧光光谱）；

26. EMC（Event Mean Concentration，事件平均浓度法）

27. Ex-P（Exchange Phosphorus，交换态磷）；

28. FA（Factor Analysis，因子分析）；

29. FCA（Fuzzy Comprehensive Assessment，模糊综合评价）；

30. Fe-P（Fe Bound Phosphorus，铁结合磷）；

31. Ful-P_o（Fulvic Bound Organic Phosphorus，富里酸结合有机磷）；

32. FT-ICR-MS（Fourier Transform Ion Cyclotron Resonance Mass Spectrometry，傅里叶变换离子回旋共振质谱法）；

33. FT-IR（Fourier Transform Infrared Spectroscopy，傅里叶变换红外光谱）；

34. GIS（Geographic Information System，地理信息系统）；

35. HPs（Hydrochemistry Parameters，水化学参数）；

36. HRU（Hydrological Response Units，水文响应单元）；

37. HSPF model（Hydrologic Simulation Program-Fortran, HSPF 模型）；

38. Hum-P_o（Humic Acid Bound Organic Phosphorus，腐殖酸结合有机磷）；

39. HUSLE（Hydro-Geomorphic Universal Soil Loss Equation，水文地貌通用土壤流失方程法）；

40. ICP-AES（Inductively Coupled Plasma-Atomic Emission Spectrometer，电感耦合等离子体原子发射光谱仪）；

41. In-stream Delivery Factor（内流输送因子）；

42. K_P（Partition Coefficient，分配系数）；

43. Kuroshio（黑潮，日本暖流）；

44. LP_o（Labile P_o，活性有机磷）；

45. Land-to-water Delivery Factor（土地-水输送因子）；

46. MLP_o（Moderately Labile P_o，中等活性有机磷）；

47. Mon-P（Monoester P，单酯磷）；

48. NLP_o（Nonlabile Po，非活性有机磷）；

49. NUFER（Nutrient Flows in Food chains, Environment and Resources use，营养流动在食物链，环境和资源利用模型）

50. OPFRs（Organo-Phosphorous Flame Retardants，有机磷阻燃剂）；

51. PO_4^{3-}（Ortho-P，正磷酸盐）；

52. PO_3^-（Meta-P，偏磷酸盐）；

53. P_i（Inorganic Phosphorus，无机磷）；

54. P_o（Organic Phosphorus，有机磷）；

55. $P_2O_7^{2-}$（Pyro-P，焦磷酸盐）；

56. Polyphosphates（聚磷酸盐）；

57. PCA（Principal Component Analysis，主成分分析）；

58. PIP（Particulate Inorganic Phosphorus，颗粒态无机磷）；

59. PLOAD model（Pollutant Loading Estimator，PLOAD 模型）；

60. PMF（Positive Matrix Factorization，正定矩阵因子分解模型）；

61. POP（Particulate Organic Phosphorus，颗粒态有机磷）；

62. PP（Particulate Phosphorus，颗粒磷）；

63. R^2（R-squared，拟合度）；

64. R_e（Relative Error，相对误差）；

65. RDA（Redundancy Analysis，冗余分析）；

66. Res-P（Residual-P，残余磷）；

67. ROS（Reactive Oxygen Species，活性氧物种）；

68. RUSLE（Revised Universal Soil Loss Equation，修正通用土壤流失方程法）；

69. SCS（Soil Conservation Service，土壤保持服务）；

70. SP（Suspended Partice，悬浮颗粒物）；

71. SPs（Socioeconomic Parameters，社会经济参数）；

72. SPARROW model（Spatially Referenced Regression on Watershed Attributes，SPARROW 模型）；

73. SRP（Soluble Reactive Phosphate，溶解性活性磷）；

74. SSA（Sewage Sludge Ashes，污水污泥灰）；

75. SWAT model（Soil and Water Assessment Tool，SWAT 模型）；

76. SWMM model（Storm Water Management Model，SWMM 模型）；

77. TDP（Total Dissolved Phosphorus，总溶解磷）；

78. Tile Drainage（暗沟排水）；

79. TN（Total Nitrogen，总氮）；

80. TP（Total Phosphorus，总磷）；

81. TPhP（Triphenylphosphine，$C_{18}H_{15}P$，三苯基磷酸）；

82. TPP（Total Particulate Phosphorus，总颗粒磷）；
83. UV-Vis（Ultraviolet and Visible Spectrophotometry，紫外-可见光谱）；
84. WTTP（Wastewater Treatment Plant，污水处理厂）；
85. XRF（X-ray Fluorescence Spectroscopy，X射线荧光光谱）；
86. XANES（X-ray Absorption Near-Edge Structure Spectroscopy，X射线吸收近边结构光谱）。

附录二 本书彩色插图

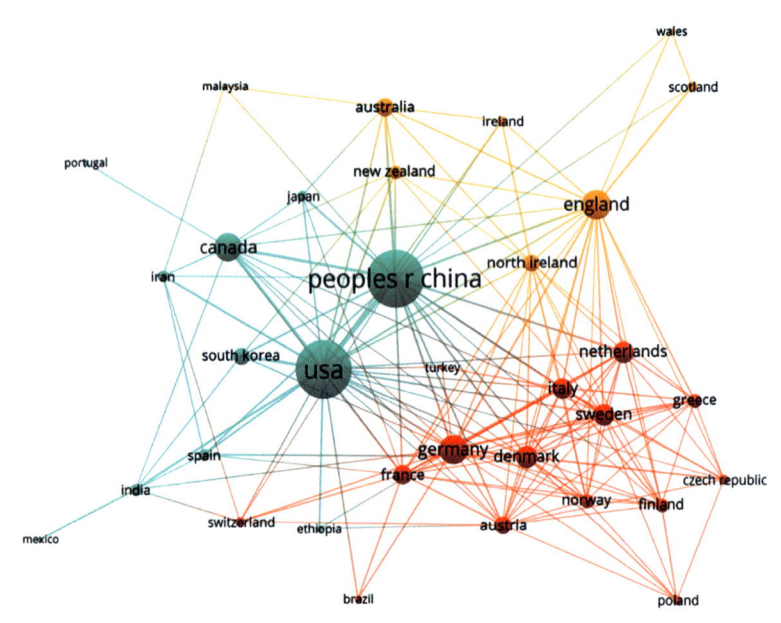

图 1-2 基于文献所属国家的文献聚类分析

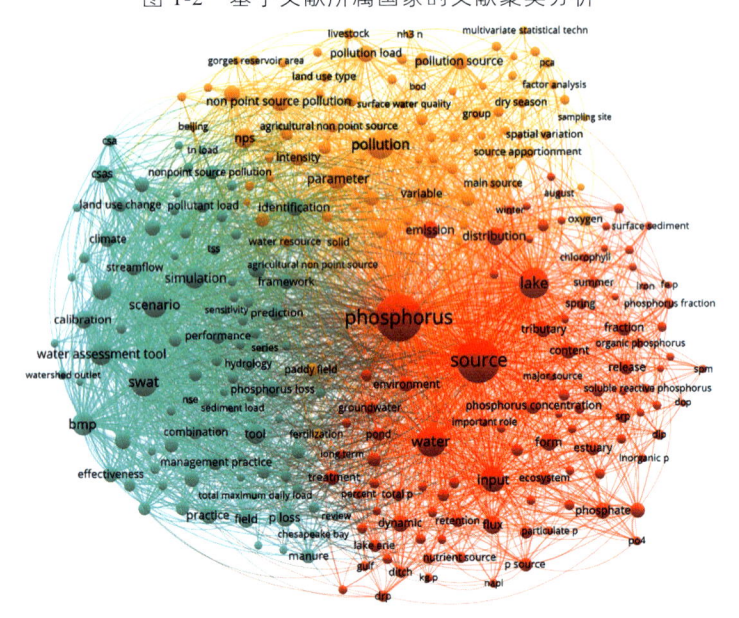

图 1-3 基于关键词的文献聚类分析

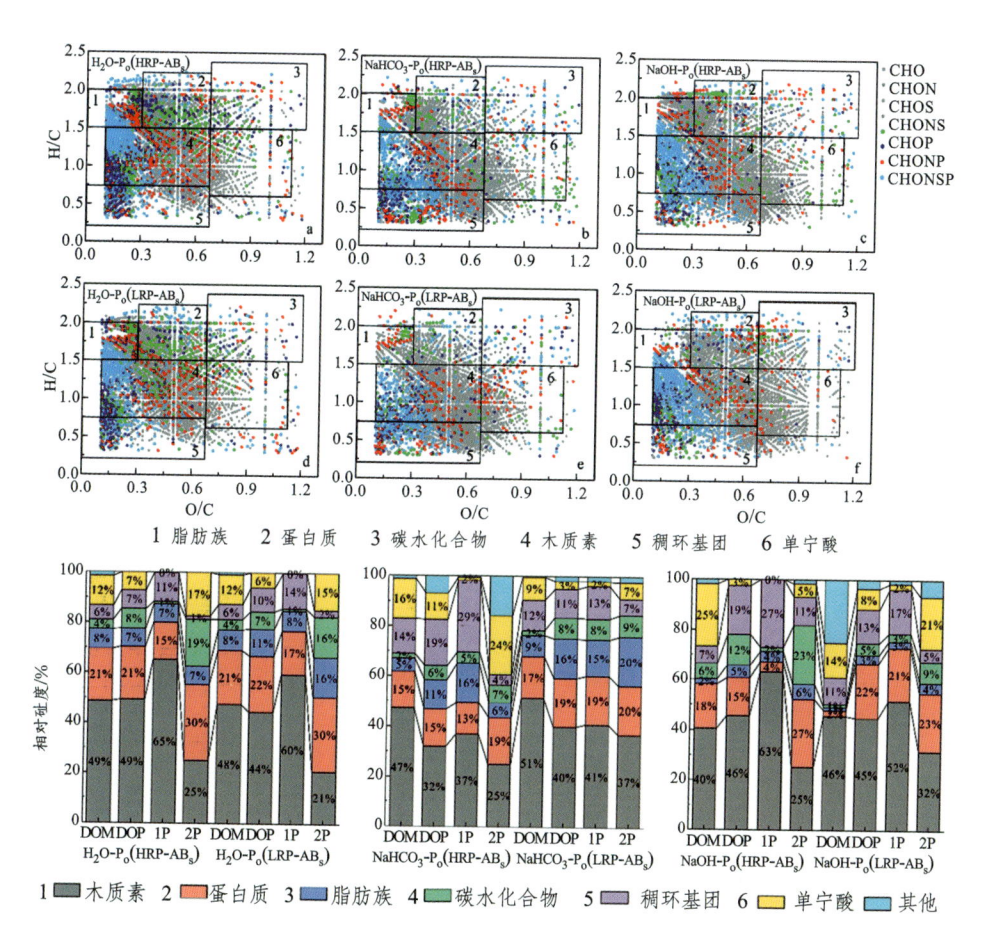

1 脂肪族　2 蛋白质　3 碳水化合物　4 木质素　5 稠环基团　6 单宁酸

1 ■木质素　2 ■蛋白质　3 ■脂肪族　4 ■碳水化合物　5 ■稠环基团　6 ■单宁酸　■其他

图 2-10　P_o 的分子组成的 Van-Krevelen 图以及相对丰度（Pu et al., 2023）

图 3-5 流域磷污染来源与流域全过程磷循环

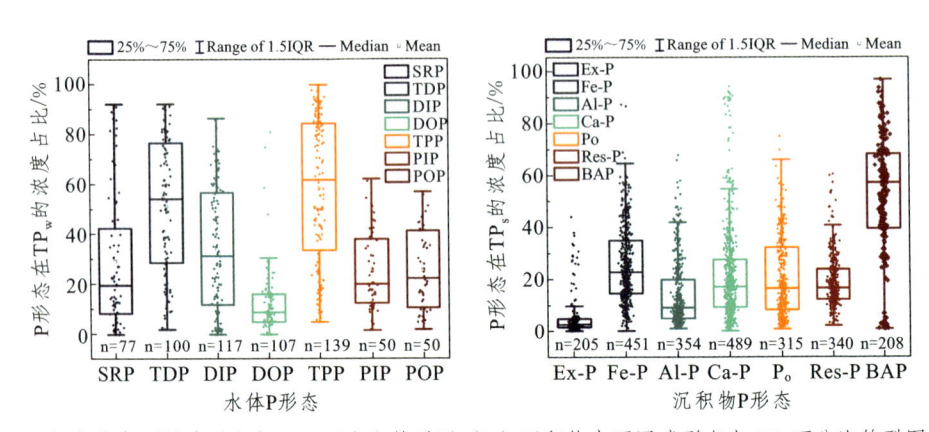

（a）水体中不同磷形态占 TP_w 百分比箱型图　（b）沉积物中不同磷形态占 TP_s 百分比箱型图

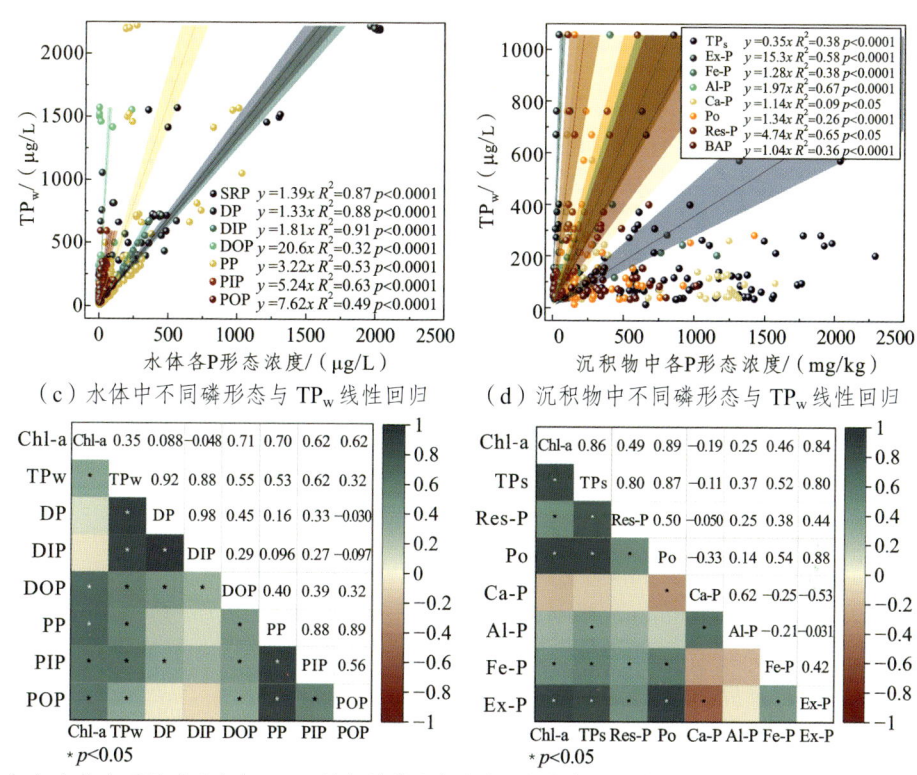

（c）水体中不同磷形态与 TP_w 线性回归　　（d）沉积物中不同磷形态与 TP_w 线性回归

（e）水体中不同磷形态与 Chl-a 的相关性分析　（f）沉积物中不同磷形态与 Chl-a 的相关性分析

图 3-6　不同磷形态对水体 TP 与 Chl-a 的影响分析

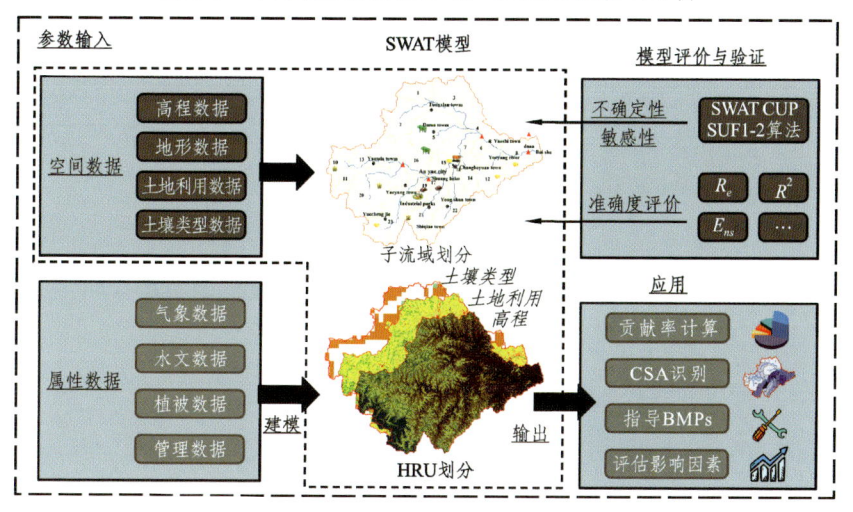

图 5-1　SWAT 建模过程，包括必要的参数输入、子流域和 HRU 划分、
模型校正以及输出结果的应用

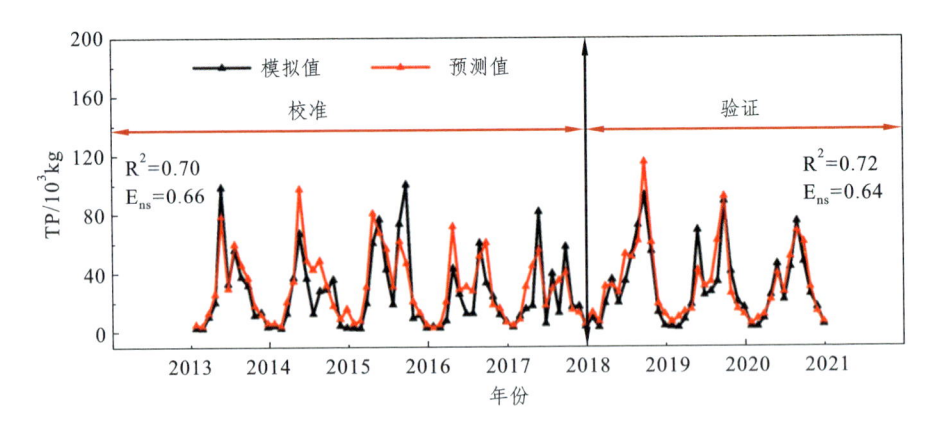

(a)模型的校准与验证

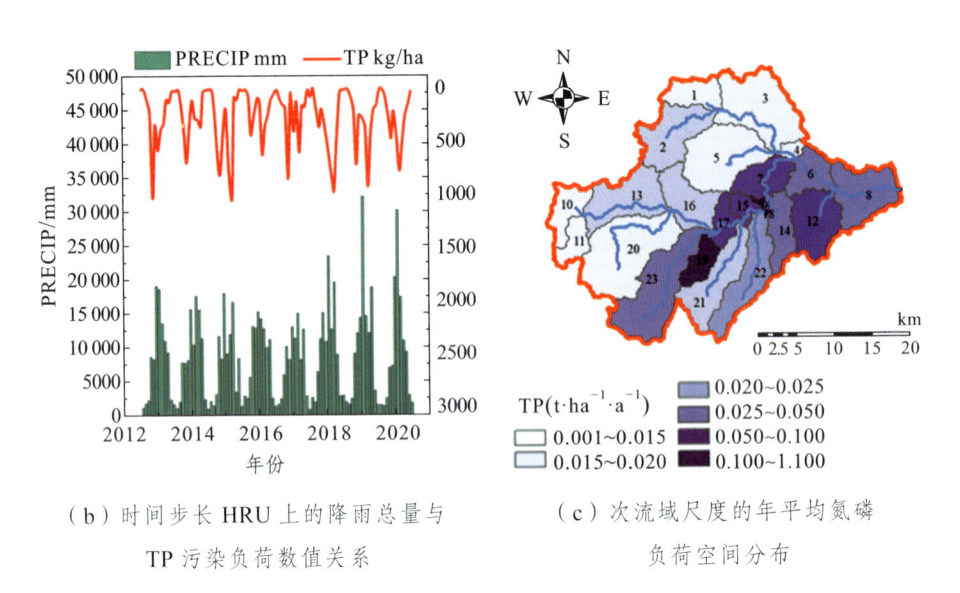

(b)时间步长 HRU 上的降雨总量与
TP 污染负荷数值关系

(c)次流域尺度的年平均氮磷
负荷空间分布

图 5-2 (a)模型的校准与验证;(b)时间步长 HRU 上的降雨总量与 TP 污染负荷
数值关系;(c)次流域尺度的年平均氮磷负荷空间分布

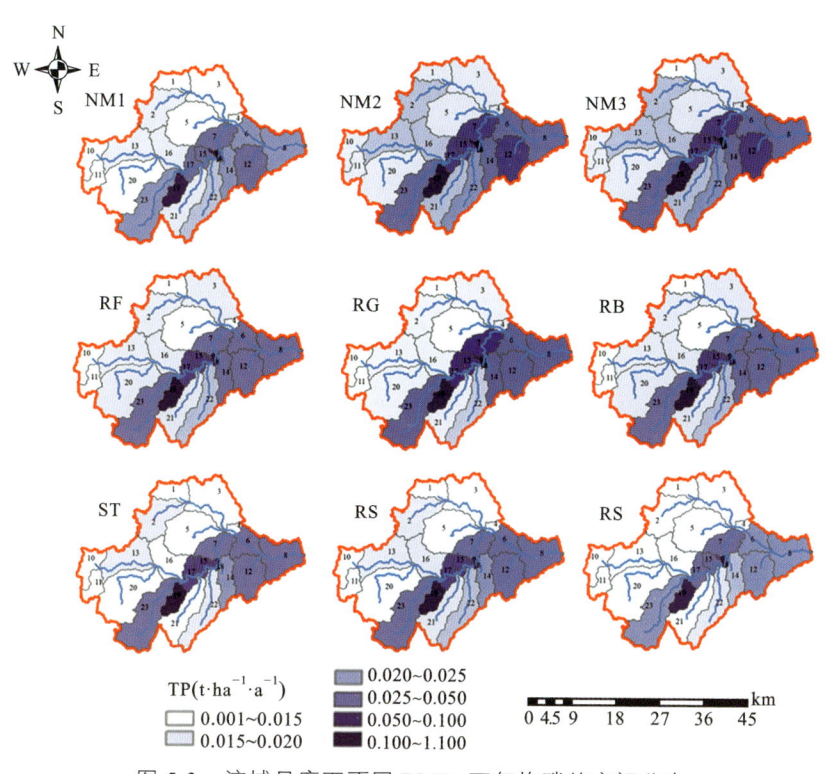

图 5-3 流域尺度下不同 BMPs 下年均磷的空间分布

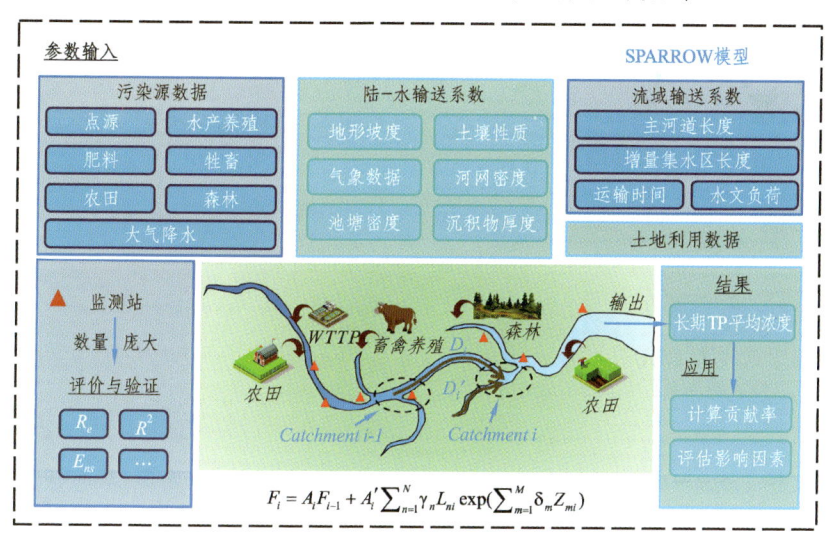

$$F_i = A_i F_{i-1} + A_i' \sum_{n=1}^{N} \gamma_n L_{ni} \exp(\sum_{m=1}^{M} \delta_m Z_{mi})$$

图 5-4 SPARROW 建模过程（包括必要的参数输入、模型校正以及输出结果的应用；除此之外图中还展示了干流增量、增量流域长度以及集水区之间的关系）

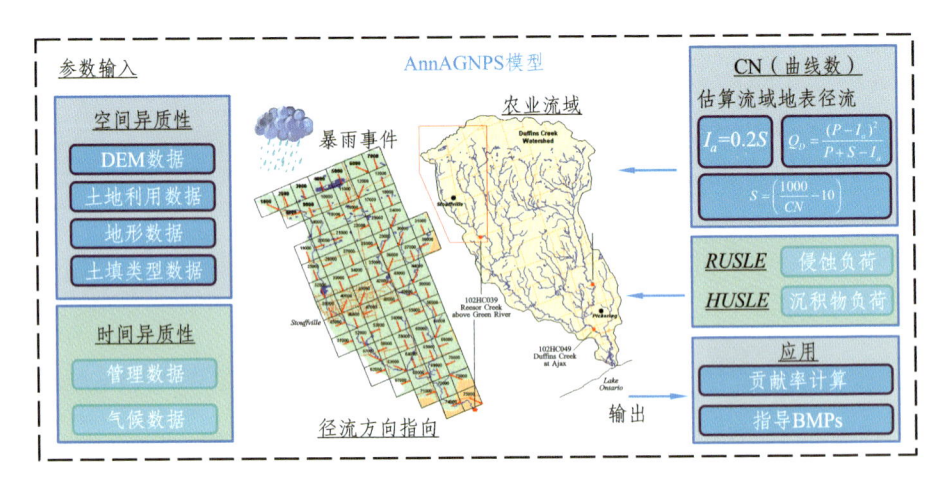

图 5-5　AGNPS/AnnAGNPS 模型的构建[图中参考了 León et al.（2004）对 AGNPS 模型的描述，以示意该模型是如何划分工作区并通过 DEM 数据模拟暴雨事件中的径流方向]

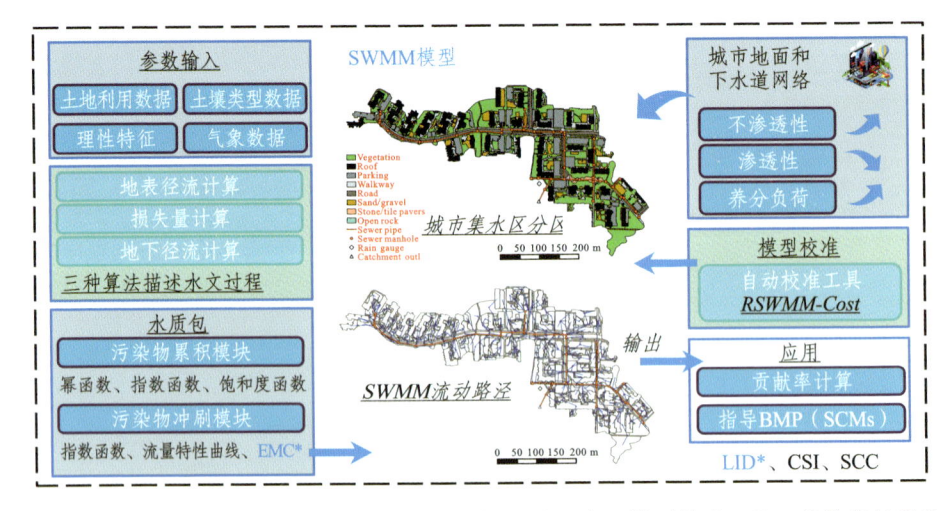

图 5-6　SWMM 建模过程[包括必要的参数输入、水质包、模型校准工具已经输出结果的应用；图中参考了 Tuomela et al.（2019）关于 SWMM 模型的描述，以示意该模型对于城市子集水区的划分以及模拟的径流和管网流动路径]

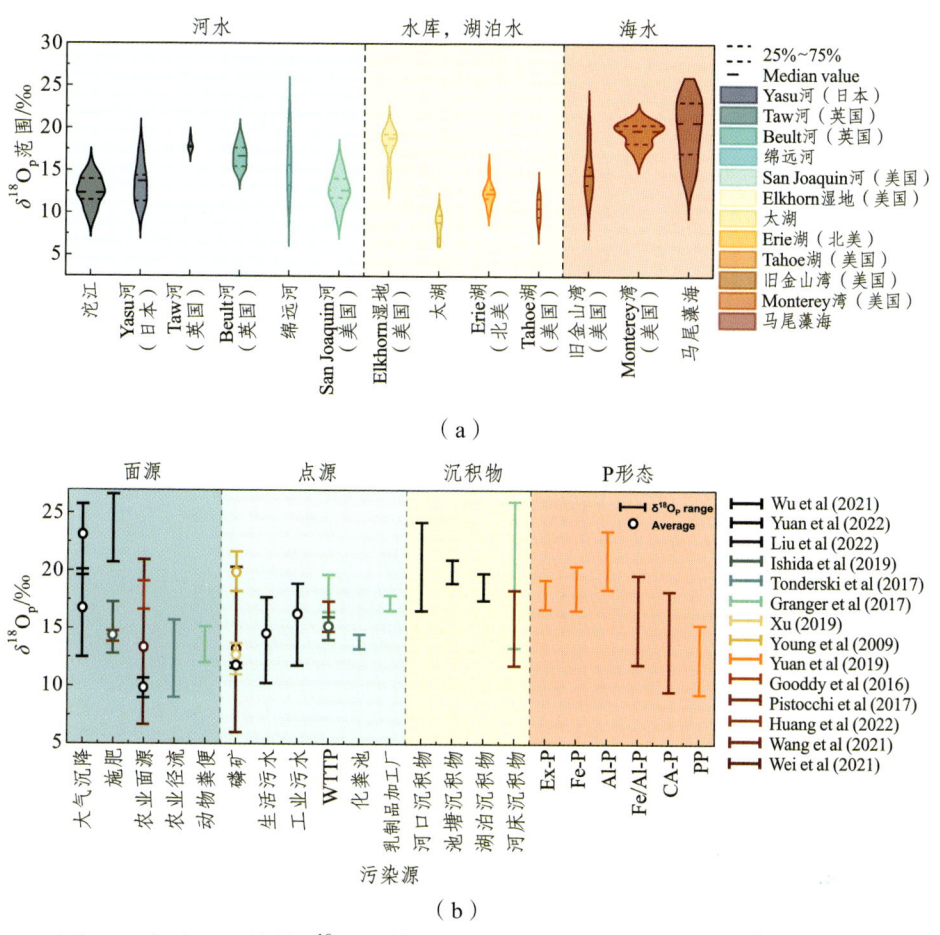

图 6-7 （a）不同流域 $\delta^{18}O_p$ 值范围小提琴图；（b）不同污染源 $\delta^{18}O_p$ 值区间

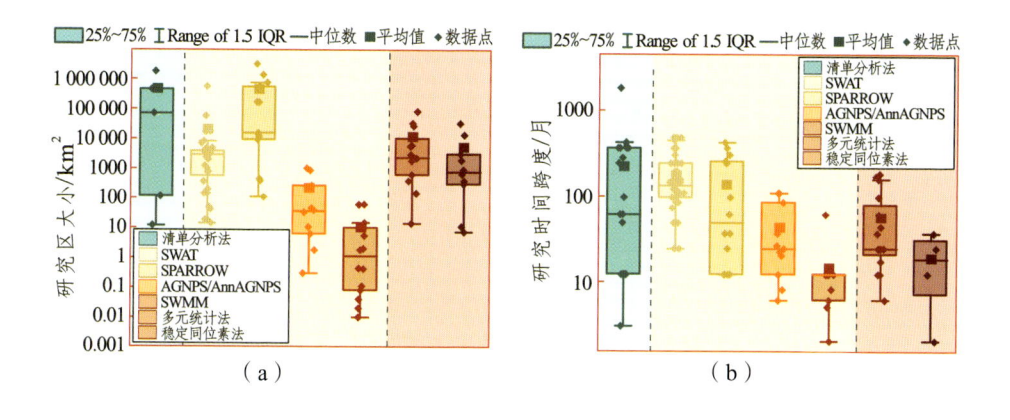

(c)

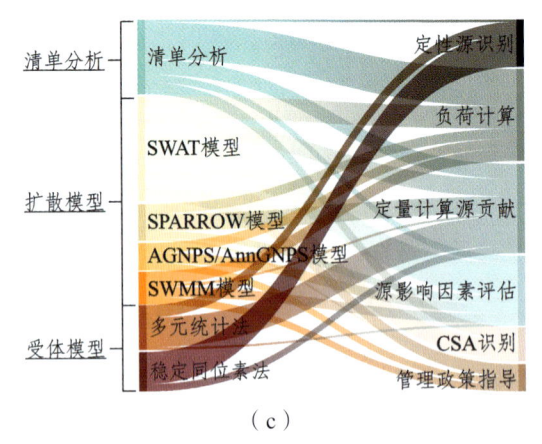

图 7-1 （a）不同溯源方法相关研究中研究区大小统计；（b）不同溯源方法相关研究中时间跨度统计；（c）不同溯源方法对应应用场景

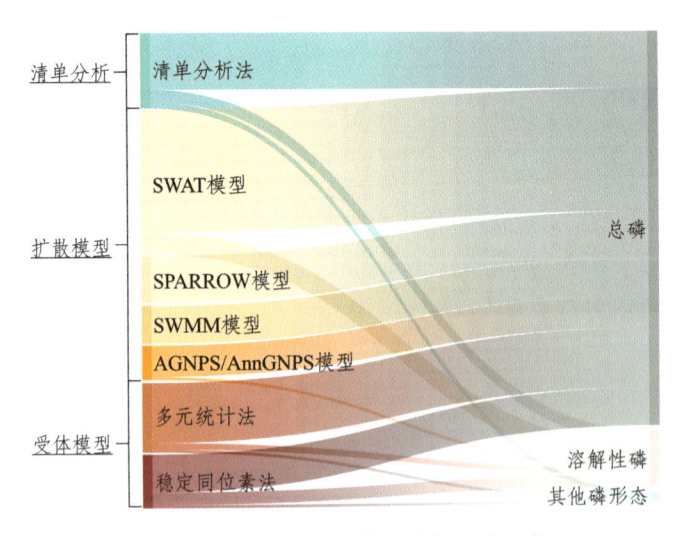

图 7-2 不同溯源方法所对应的磷形态